# TRANSPORT DESIGN

MRB-301

# TRANSPORT DESIGN

## A Travel History

Gregory Votolato

REAKTION BOOKS

Published by Reaktion Books Ltd
33 Great Sutton Street
London EC1V 0DX, UK

www.reaktionbooks.co.uk

First published 2007

Printed and bound by Cromwell Press Ltd, Trowbridge, Wiltshire

*British Library Cataloguing in Publication Data*
Votolato, Gregory
    Transport design : a travel history
    1. Vehicles - Design and construction  2. Human engineering
    I. Title
    629'.046

    ISBN-13: 978 1 86189 329 1
    ISBN-10: 1 86189 329 9

# CONTENTS

# INTRODUCTION

## Modern Travel

All vehicles are conceived as temporary accommodation for the traveller, but each is devised and equipped for journeys of particular lengths. Hotel-like ocean liners and mechanistic orbital space stations can accommodate their occupants for weeks or months, while today's city buses, in which most passengers stand, are normally occupied for minutes rather than hours. In the former case, entrances and exits are typically inconspicuous, while passengers on a bus are constantly aware of the doors and often want to remain near them. Time is one of the primary factors in transport design.

Speed is another. In modern jet fighter planes the pilot's visual reference to the world outside the cockpit, even at low altitude, is unreliable because of the plane's velocity. Digital instrumentation is the key to successful missions, and infrared night-vision is today's ultimate instrument in the delivery of 'smart' bombs to destroy targets that the pilot cannot see with the naked eye. Yet to many people slowness is still a good thing, which is one reason why sailing remains a popular leisure pursuit, providing an intimate connection with the natural world and a benign exploitation of nature's resources for the pure pleasure of personal mobility. Designs of even the most advanced modern sailboat cockpits retain the essential features that connect the sailor directly with sun, wind and water.

Like buildings, some vehicles are designed from the inside out, while others are conceived with a predetermined form into which people and activities must fit. Many modern cars are shaped around the ergonomics of the passenger cell. Alternatively, if office buildings are often shaped like cereal packages, certain types of vehicles are either boxes or tubes, whose forms are conditioned by factors other than the needs of their potential occupants. Such factors include economics, construction technology, laws of physics or a particular infrastructure – street layout in the case of architecture, road or rail networks in relation to vehicles. Nevertheless, although they are similar in shape, the difference between

The panoply of early 20th-century transportation, from a 1910 lithograph.

7

ONE OF THE 28 *EMPIRE* FLYING - BOATS
TWO DECKS . 200 MILES-AN-HOUR . 18 TONS

IMPERIAL AIRWAYS

The curvaceous fuselage of the Short c-class Empire Flying Boat created challenges for the interior designer Brian O'Rorke, who created rectilinear, room-like cabins within the organic shell of these ships, which entered commercial service in 1937.

the interior of a railroad boxcar and a Pullman dining car is fundamental, just as a house and a warehouse are different species.

A relation between vehicles and architecture is fundamental to this study, since the systematic human understanding of *place* is dependent on similar principles whether one is in a building or onboard a moving vehicle. The body of any conveyance, like the shell of a house, defines our interface with the vastness of space beyond the enclosure. Since antiquity, the discipline of geometry has given us an intellectual grasp of the spaces we inhabit and of those beyond our immediate field of vision.

The space–time concept central to Einstein's theory of relativity carried us further towards an abstract understanding of our place in the universe, but it contributed little to our personal need for orientation. Yet throughout the history of transportation, builders of ships, aircraft and automobiles have attempted to create interior spaces that communicate a sense of stability and place. The modern study of psychology has also contributed to a greater understanding of how we perceive space, recognizing the complexity and variability of human experience. For example, a driver and a pedestrian respond differently to the same streetscape, their actions based on considerations specific to the role of each at the time.

At a personal level, our notions of space are derived from direct experiences supported

Boxcar: the Austin Seven of 1922, seen in sectional view, represents a type-form, the rational, architectural vehicle, its engine and passenger compartment fitted into a two-box form.

THE AUSTIN SEVEN SALOON
IN SECTION

by abstract learning. We learn the pattern of the solar system, although few of us have ventured beyond the gravity of the Earth. Map reading is a common abstract skill that enables us to find our way around city and countryside. Alternatively, children learn about the spaces within their homes by direct, tactile experience, and their perceptions of those spaces change as they grow. When we travel in a vehicle the same cognitive processes condition our knowledge of the space within that vehicle and our understanding of the space outside it.

If our prior conditioning determines the way we experience travel in a vehicle, our experience is also informed by the design of the vehicle's interior; and that is the subject of this book. During the past 200 years all types of vehicles and vessels have evolved relentlessly due to the desires of their designers and manufacturers (and the demands of their users) to improve their performance. By performance, I mean the technical efficiency with which they function, the extent to which they satisfy the various practical and emotional demands of their owners, operators and users, their commercial success, their longevity, and the esteem in which they are held in relation to their competitors.

Beyond moving us from A to B, every vehicle in which people travel provides an interface with the natural, physical world. The structure of an aeroplane, ship or car can be seen as an extension of the human body, an artificial carapace that buffers those inside against the effects of speed, changing air pressure, extremes of heat and cold, the unevenness of surfaces over which we move, turbulent air and sea, or the vacuum of outer space.

The space station settings for Stanley Kubrick's film *2001* caricatured the sleek, organic minimalism of 1960s Italian furniture design, set in an abstract, minimalist interior.

These same vehicles also provide us with community and privacy. For 150 years the grand, nautically engineered and architecturally decorated spaces on board ocean liners have been the venues for a highly contrived social experience revolving round formal dinners, the ritual promenade or a game of shuffleboard with a fleeting acquaintance. The interiors of our cherished cars protect us from the random human contacts inflicted by public transport, while the extensively equipped sleeper pods in upper-class sections of jumbo

jets enable thrusting executives to subside into peaceful oblivion at 35,000 feet, without making eye contact or conversation with fellow travellers accommodated only a few inches away.

Yet some features of our best-loved vehicles can quickly become hindrances. A comfortable inertia-reel seat belt will stop a motorist from rocketing through the windscreen in a frontal collision, but becomes annoying when it wrinkles our clothes. More disastrously, the trusted airbag has caused the deaths of children and smaller adults seated in front during even relatively minor car crashes. Advanced technology has often brought new dangers. Thirty-six passengers perished in the dramatically modernist lounges and staterooms of the impressively engineered zeppelin LZ129, the *Hindenburg*, as the ship's fabric covering and hydrogen-filled gas bags exploded in flames, and the fire-swept nacelle quickly crashed to the ground at the aerodrome of Lakehurst, New Jersey in 1937. By the end of the 1930s, the invention of pressurized aircraft cabins enabled us for the first time to fly high over the worst effects of the weather. Yet sixty years later the American golf champion Payne Stewart and his retinue died bizarrely in their private Lear jet when the cabin suddenly depressurized at 30,000 feet. The plane and its macabre cargo of corpses then flew on autopilot, its windows ominously frosted over, for four hours before running out of fuel and crashing.

In the twentieth century mechanized mass transportation opened up exciting new experiences for people of all classes. Yet boredom was an ever-present threat in modern travel. Consequently, the interiors of pre-Second World War trains, aircraft and ships, at least for the upper classes, were designed as havens of escapism and luxury, employing every

Privacy at bedtime has always been the Holy Grail of transport interiors. Tangerine Design devised these opposite-facing sleeper pods for business-class passengers on British Airways long-haul flights (2001). High equipment specification and movable screens enabled business fliers to work or relax with minimal intrusion.

known strategy for reducing tedium and anxiety. To this end, from the mid-nineteenth century, shipboard cocktail parties and seven-course dinners were staged elaborately amidst the diverting interior decor of cafés, bars, ballrooms and other hospitality spaces aboard the world's great ocean liners and cruise ships.

Early airlines used the latest entertainment technologies to distract their passengers. By the mid-1920s the pioneering British airline Imperial Airways had shown the first in-flight movies using hand-cranked projectors accompanied by the music of wind-up phonographs, an innovation in passenger entertainment put on hold for many years following the arrival in 1927 of sound films – early 'talkies' required heavy equipment, and aircraft cabins were too noisy for the dialogue to be heard.

Today, the boredom of a daily commute to work by car can be relieved or even raised to the level of a private pleasure through the technology of digital music systems located in the dashboard or controlled from the steering wheel. Although the pleasure drive may be only a memory in the traffic-congested twenty-first century, many motorists can still identify with the lyrics of Chuck Berry's classic rock and roll 'driving song' of 1964:

> Riding along in my automobile
> My baby beside me at the wheel
> I stole a kiss at the turn of a mile
> My curiosity running wild
> Cruisin' and playin' the radio
> With no particular place to go

The technology of in-car entertainment has advanced from early, single-band valve radios to the sophisticated multi-functional stereophonic CD/radios installed in most modern cars. Yet the urge to be serenaded while driving has remained strong from the start – music, romance and the automobile. Freudians describe the irrational love we feel for these inanimate objects as 'cathexis'.

The entertainment function in cars has been developed further thanks to the introduction of video technology, with screens built into the backs of the forward headrests for the benefit of rear-seat passengers on longer journeys, just as in an airliner. Airlines, themselves, reintroduced film screenings on long-haul flights in the 1960s, using the tubular space of the fuselage as a multi-screen cinema. In the era of wide-bodied jets with a small proportion of window seats, the personal 5-inch video screen has replaced the sense of spectacle formerly provided by the airborne cinema or by window views of landscape and cloud formations.

The interiors of all kinds of vehicles have long been designed for the appreciation of spectacle, that astonishing, awe-inspiring sight, sometimes enhanced by the element of risk. The specific spatial qualities of vessels and vehicles can create amazing views, as in the linear promenade decks of traditional ocean liners. The fenestration of each type of vehicle frames images of the world outside in particular ways – panoramic views from modern train carriages and coaches or focused views of the Earth's curvature seen through the tiny windows of the supersonic airliner, Concorde, flying at an altitude of 60,000 feet.

In addition to the traveller's-eye view of the passing scene, communication devices, such as cellular telephones and in-car satellite-navigational systems, offer even the most hopelessly inept map-reader the information necessary to reach almost any destination. They also offer diversion and amusement through their game-like qualities. These relatively inconspicuous black glass devices, found in all sorts of vehicles, are magically transformed by the colourful pictorial realism of the software, which can show building contours of unfamiliar cities, bird's-eye overviews of roadways or harbour channels, and significant features of natural terrain to clarify and bring to life what in the recent past would have been seen as abstract, diagrammatic information. Thus, the visual quality of the video game has come to the driver's, skipper's or pilot's seat. At the same time, the equipment can be used to download music from an iPod or run DVDs for kids, to eliminate the chant commonly heard on long family drives: 'are we there yet?'

Modern transport interiors have ranged from deeply drab to intensely glamorous. Design has been employed to provide the most basic, utilitarian environments in which masses of people have moved within cities and across the globe. Buses, ferries, trains and aircraft have served the purposes of daily commuting, troop transportation and mass migration. Some public vehicles, such as the London taxi, the rubber-tyred carriages of the Paris Métro and the Douglas c-47 aerial troop carrier, have all earned the respect of those who used them, regardless of their relatively spartan interior comforts. At the other end of the spectrum, vast amounts of money, design talent and craftsmanship have been employed to cosset the princesses and playboys who rode in Bugatti Royale limousines before the Second World War or the billionaire owners of today's Pershing or Ferretti yachts docked in marinas from Monaco to Miami.

The difference between the haves and have-nots in society has never been more starkly revealed than in the design of transport vehicles, for example in the first-, second- and third-class accommodations aboard the steamships that provided the main form of international transporta-

The Handley Page HP.42 airliner was already a dinosaur when it went into service with Imperial Airways in 1934. Its carefully detailed and luxurious interiors were, however, very popular for their spaciousness, large windows and Odeon-style decoration. Onboard service was prodigious.

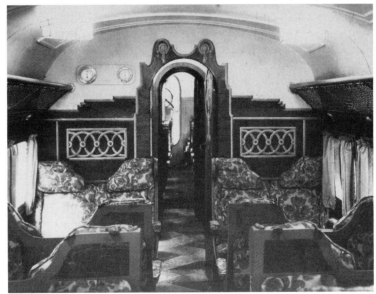

tion from the middle of the nineteenth century until the development of jet airliners. On board these leviathans, the wealthy enjoyed palatial surroundings carefully designed to confirm their sense of importance and superiority. Meanwhile, the poor survived their voyages in quarters designed for maximum efficiency and minimum amenity.

The specific character of each type of transport interior was in almost every case slow to develop. Designers, owners and builders of new types of transport vehicles relied on precedents in earlier kinds of vehicles or in the familiar terrain of architectural interiors. In the early days of air travel, for instance, some aircraft interiors were conceived as airborne Pullman cars, employing the internal layout, seating design and decor familiar in elite railroad carriages such as the Orient Express. Alternatively, Hugo Junkers, the designer and manufacturer of the first all-metal airliner, the F 13 of 1919, described that plane as a 'flying limousine'. Its interior, seating four on an upholstered bench and two individual 'bucket' seats, could be easily mistaken for the passenger compartment of a contemporary Mercedes or Daimler, complete with a partition between the passengers seated inside the cabin and the pilot, exposed to the elements in an open cockpit. Only in the 1930s did the tubular interior of the airliner develop an aesthetic of its own, related to its specific materials, construction, function and spatial qualities.

The aesthetic development of some transport interiors followed the fields of architecture, furniture design or other established disciplines in relation to style. In the nineteenth century many of the royal trains of Europe were designed with interiors that emulated the grandest rooms of the monarchs' palaces. Others were designed and constructed in the manner of royal coaches, the appearance of their interiors heavily dependent on the upholsterer's craft.

Conversely, many innovations conceived specifically for transportation influenced the styling of conventional furniture. The tubular steel bicycle frame is often cited as the source of inspiration for lightweight metal furniture designed at the Bauhaus in the 1920s. The technical glamour of the adjustable rail seat and the lightweight airline chair also rubbed off on domestic and commercial furnishing and architectural interiors. Streamlining was conceived as a practical means of increasing the efficiency and speed of trains, aircraft, ships and automobiles, yet the sleek curvaceous external forms of those vehicles and vessels transferred readily to interior design and furniture. Kem Weber's cantilevered timber Airline Chair, designed in 1934 for the Walt Disney Studio, demonstrated the popular application of streamlining as an interior design fashion.

Since the 1960s, when health and safety were finally becoming established as moral rights for anyone who travels, engineering, styling, manufacturing and marketing have made them increasingly dominant factors in transport design. Even earlier, in the 1920s, the airline watchdog and regulator, IATA, was formed to develop and ensure international air traffic control mechanisms, but also to establish basic standards of

In the developing world, passenger safety often remains a low priority in relation to basic needs of transportation. Rooftop riding on public buses and trucks is reminiscent of the nineteenth-century knifeboard omnibus, precursor to the double-decker bus. However, the elaborate ornamentation of this Pakistani bus lifts it well above the banal appearance of public buses in North America and Europe.

aircraft seat design, space provisions and sanitation. Regulation for designed-in safety came slower to our roads.

The American automobile industry was rocked by a series of highly publicized Congressional hearings during the 1960s, during which automotive interior design was severely criticized for failing to consider passenger safety as a priority. Since then, manufacturers across the world have invested increasingly in the provision of sophisticated seat belts, more defensively structured passenger compartments, multiple airbags and many other innovative devices to protect those inside cars during an accident. Coincidentally, they also improved the air conditioning, seat controls and other comfort devices that help to keep drivers alert in all conditions.

Improved design standards, however, have not always been able to prevent accidents, and legislation has not always been able to impose sensible measures to reduce deaths and injuries. Seat belts are still not fitted universally in motor coaches, which travel at high speeds on the world's motorways. Meanwhile, in developing countries passengers are still riding on the rooftops or hanging from the sides of overcrowded buses and trucks. In those same parts of the world, in spite of good design, ferry disasters are frequent and casualties high due to overloading, poor maintenance and the casual enforcement of basic safety regulations.

With a growing awareness of environmental issues around the world since the 1960s, the design of vehicles has been subjected increasingly to scrutiny over their contributions to global warming and their other deleterious effects on the natural habitat. Noise, exhaust emissions, fuel consumption and manufacturing processes are only some of the dubious aspects of modern vehicles. Their production alone involves the mining of metals, the manufacturing of plastics and many other energy-greedy processes that pollute or exploit the land and the atmosphere.

The very seductive nature of these mobile environments is a crucial factor in encouraging modes of travel that may become unsustainable in the foreseeable future. The much-loved private car interior is a prime example, but so is the economy section of the modern, high-density airliner cabin. While the desire to travel may be a natural human urge, interior design, like advertising, is a powerful instrument in the commercial promotion and realization of that urge. Recycling, design for disassembly, adaptive reuse, the development of new and more ecologically friendly materials, and the promotion of the more efficient transport types, such as the train, all aim to reduce the ill effects of modern travel.

Throughout the last century many utopian proposals for innovative transport designs were published or developed to prototype stage. These ranged from giant, flying ocean liners to automobiles that drove themselves. Many were dead ends, but some promoted ideas that would become fundamental to later commercial designs. The interiors of the Cornell-Liberty Safety car of 1958 illustrated the concept of passive safety. This 'ideas' vehicle, produced by the Cornell University aeronautical laboratory in collaboration with the Liberty Insurance Company, previewed advanced passenger-restraint mechanisms, fully padded, impact-absorbing interior surfaces, burst-resistant door locks and a seating layout developed to improve the driver's view and control, a layout that has been employed in the design of modern cars as unique as the McLaren F1 and the Smart city runabout.

Some of the most influential ideas about the designed environments in which we travel have emerged from science fiction, such as the novels of Jules Verne and H. G. Wells; and these have sometimes overlapped with the work of renowned twentieth-century designers, such as Norman Bel Geddes and Buckminster Fuller. Some of the latest visionary designs also hark back to historical notions of travel retained in the popular consciousness. Prototypes of the Airbus A380 super-jumbo airliner feature lounges, promenades and other interior features reminiscent of 1930s flying boats and other luxury transport from the pre-jet age. Whether these niceties will be realized in commercial use remains questionable, but the urge for luxurious mass transportation remains strong, even in today's harsh travel economy.

At present, such luxuries are enjoyed mainly by the lucky few travelling in private aircraft, yachts, cruise liners, nostalgia trains such as the

The Airbus A380 is the first of a new generation of double-deck jumbo jetliners promising more generous internal spaces, which could enable airlines to provide amenities such as lounges and therapy rooms, offering passengers opportunities to leave their seats during long-haul flights.

Trans-Siberian Express or corporate-owned Maybach limousines. For the rest of us, such experiences come vicariously through the movies or the fairground. Disneyland's Test Track in their Florida Epcot Center is a case in point, where visitors can experience a futuristic vision of private transportation sponsored by General Motors.

The following three chapters cover time scales beginning with the first major developments in the modes of transportation being discussed. The chapter on land vehicles begins with the development of the horse-drawn carriage in the late eighteenth century, when city traffic was growing and country-wide road networks were first being developed. The chapter on ships and boats begins in the nineteenth century with the earliest steam-powered trans-oceanic passenger liners. Coverage of air travel starts with the Wright Brothers at the beginning of the last century. Yet the order of each chapter is primarily thematic, rather than chronological, revealing links between developments across different modes of transport and over time. Finally, we will look at how events unfolding in today's turbulent world are shaping the next developments in transportation design.

Habitability became a buzzword for the interior design of NASA's Skylab orbital space vehicles during the 1970s, since astronauts would be living and working for months in these mechanistic environments. Raymond Loewy considered his work on Skylab and its associated shuttle vehicles to be the most important project of his career.

During the past 200 years the ways in which people around the world have travelled, locally and globally, have been increasingly defined by the nature of

The 'tub' or cockpit of racing cars is typically an austere affair. From the earlist racing cars of the 1890s, to today's Ferrari Formula One cars, these cockpits have demanded of their drivers the highest skills and bravery and entertained the greatest thrills of any land vehicles.

Even in the age of mass air transportation, the view from an economy-class aircraft seat can dazzle and inspire the most jaded traveller. View from a Boeing 737 over the Pyrenees.

their work and by the way they spend their leisure hours. At the same time, those corporations and governments that control most aspects of our transport networks, our vehicles, aircraft and vessels, have exploited both design and publicity to compete in heavily saturated markets. Alongside developments in technology, design became an essential tool in improving the habitability of transport vehicles. It also transformed many practical conveyances into mobile people's palaces that could be experienced by nearly everyone in the industrialized world. Inevitably, these modern magic carpets became the subjects of writers and visual artists and the stuff of popular mythology.

The media feed us images of troubled film stars and disgraced politicians hiding behind darkened glass in the richly upholstered passenger compartments of chauffeured limousines. Televised diaries show

US astronauts floating weightless in the busy mechanical interiors of their orbital capsules, shuttle vehicles and space stations. Formula One cockpit videos and video games put us in the driving seat of F1 cars to simulate the thrilling adrenalin-fuelled racing exploits of Ralf Schumacher or Juan Pablo Montoya. Meanwhile, in the world of every-day experience, millions of travellers in the cramped economy-class seats of jumbo jets watch with awe the stunning beauty of the earth as seen from 35,000 feet.

With an increasingly mobile world population, the interiors of vehi-cles will be subjected to ever greater levels of public and professional scrutiny, technical innovation, governmental or international regulation, and aesthetic enhancement to protect, seduce and entertain us in our peripatetic lives. The following story attempts to explain how and why these familiar or exotic conveyances have come to be as they are and how they affect our travel experiences.

# 1 LAND

## Kings of the Road

During the later eighteenth century, in the newly industrializing, commercially minded and imperialistic countries of Enlightenment Europe, a relentless quest arose for improved communication. A new network of relatively well-maintained roads spread from the major cities out to the provinces. By the 1820s France could boast of nearly 16,000 miles of surfaced roads, some up to 13 metres wide, allowing the easy passage of large vehicles drawn by teams of four or six horses running at a steady gallop. Thus, up to the invention, growth and ultimate supremacy of the railways, there arose an important period in the development of horse-drawn road vehicles. This was the start of the age of speed in which we live today.

Large elaborate carriages were used in particular by those who regularly travelled long distances: diplomats, royal courtiers, the higher nobility and artists on their Grand Tour. Typically, the wooden-framed bodies of these carriages were suspended above a Berlin-type undercarriage by steel c-springs for comfort and stability. The best springs and thus the most comfortable carriages were made in England, which in the eighteenth century had the most advanced metallurgical industry. Typically, the driver sat on a flat board perched above the front wheels, while the passengers occupied a compartment, which had a u-shaped profile and a flat roof above.

Many of these heavy vehicles were elaborately ornamented, prototypes of what has been called 'hollow, rolling sculpture', and they were decorated by some of the most important artists of the time. Royal Academicians such as Sir Robert Smirke, architect of the British Museum, Charles Carton and John Baker all painted decorative panels, floral borders and heraldry for carriages. In addition to Berlin coaches (named after the city where this type was first built), the English travelling carriage became a popular vehicle built according to widely published plans by many carriage-makers around the world. The Italian soldier, adventurer, diarist and seducer Giacomo Casanova owned several, the

Aerial view SE along Virginia Avenue, Washington, DC, c. 1992.

21

best-documented of them built by Hofsattlerei of Vienna in the 1770s; and he used them for more than simple transportation. Love and sex also took to the road.

Travelling carriages would be drawn by a team of several horses and typically accommodated four to six passengers luxuriously in a closed compartment with large glazed windows (which could also be shuttered or curtained for privacy) to admit plenty of light into the cabin, to frame the passing views of landscape and to allow the

occupants to show themselves off to those outside. Although their forms varied considerably, many travelling coaches were fitted with amenities otherwise found only in the best houses.

In an era before hotels, the interior of Cardinal Richelieu's carriage, *c.* 1625, could convert from a mobile parlour in the daytime to accommodate a full-sized bed, heated and made up with feather pillows and eiderdown. The Berlin in which the painter Jean-Honoré Fragonard and his wife travelled from Paris to Rome in 1773–4 was equipped for elegant dining, when the coach would stop for meals and to rest the horses. Its fitted cabinetry was designed to carry the artist's drawing folios and his library of books, and the interior was richly upholstered, curtained and carpeted. The Fragonards' journey lasted for two months, during which much of their time would have been spent inside the carriage.

For himself and his retinue, Napoleon I employed a number of sleeper Berlins, with interiors of green or blue satin with red and gold trimming. These were equipped with a writing desk and a chandelier – to ensure that work could proceed uninterrupted while the emperor and his retinue travelled – and other amenities, including a clock to mind the speed of his progress. A cooking stove, wine cellar, provisions for palace-quality dinners and folding campaign furniture, on which he would dine, were transported in other carriages of his convoy also occupied by the cook and his staff. Speed was of the essence. One of his aides noted that, upon arrival at the 'field kitchen' wherever the emperor decided to camp for dinner,

> Napoleon takes his meal (and the speed with which he eats is common knowledge) in company with his senior officers . . . and then

An archetype of 'hollow rolling sculpture', this coach was built in 1757 by Joseph Berry as the Lord Mayor of London's ceremonial carriage. It became an enduring symbol of the English capital, yet its Rococo sculptural decoration was based on a style imported from the French court of Louis XV and its elaborate decorative paintings were executed by the Florentine artist Giovanni Battista Cipriani.

everything is rapidly stowed away; the provisions, the crockery, the silver; and the journey continues.[1]

This is a distant prototype for the mobile offices in the chauffeured cars of countless business executives on the move in our modern commercial world. Napoleon had the capability to keep working while moving at the highest speed that the technology of his time allowed, just as today business people are compelled by their work ethic to fax, text, email and conference call while cruising along the motorway at 100 miles per hour between appointments.

The emperor's vehicles were, of course, unique, yet travellers' accounts and engravings of the time demonstrate the large number of passenger vehicles on the streets of European cities during the later eighteenth century. One visitor to Warsaw in the 1770s commented wryly: 'Here, no one from the average shopkeeper upwards would dream of walking.'[2] In fact, Paris alone counted some 14,000 public and private vehicles by 1763, Madrid between 4,000 and 5,000 in 1780, and Vienna 3,000 at the same time. The European aristocracy, who owned or had access to carriages, used these vehicles as highly visible symbols of their importance in the community and the court.

The Royal Mews at Buckingham Palace in London houses a collection of coaches ranging from the Gold State Coach, built for King George III in 1762, to the Australian State Coach, constructed as a gift for Queen Elizabeth II in 1988. The former is an archetypal Cinderella carriage, its design supervised by the prominent architect Sir William Chambers. The Italian artist Giovanni Battista Cipriani, who also decorated the Lord Mayor of London's coach, painted the exterior panels and its gilded wooden body, weighing 4 tonnes, is ornamented with carved Tritons and palm trees. The plush interior is upholstered in crimson silk, tufted and buttoned over the seats and wall linings. It provides upholstered armrests and embroidered sashes to raise and lower the windows.

In modern times the coach's interior has been modified to enhance its function as a dramatic stage prop. Internal theatre lighting, added for Elizabeth II's Coronation in 1953, was colour balanced so that the queen would appear at all times as if in full daylight. Special fittings were also provided to support the royal orb and sceptre, which are too heavy for the queen to hold throughout a long procession. Yet a hot-water bottle provides the only heat. In contrast, the Australian State Coach, despite its traditional appearance, with blue silk brocade upholstery and Waterford Crystal lamps, has electric windows and air conditioning, powered by an on-board generator, for the comfort and convenience of its occupants during formal processions such as the State Opening of Parliament and royal weddings.

Although royal carriages retain their unique if anachronistic ceremonial role in contemporary culture, as Thorstein Veblen declared in his *Theory of the Leisure Class* (1899), in the nineteenth century the private carriage was, along with the city mansion, one of the most visible symbols of private power and wealth. Their elegant lines, gleaming surfaces and sumptuously appointed interiors provided the setting in which the privileged went out to play and were seen to play. One of the most successful types was the brougham, named after its designer, the Englishman Lord Henry Brougham, who commissioned the first model in 1838–9 for his personal use.

The first four-wheeled closed carriage light enough to be pulled by a single horse, the brougham was designed to be coachman-driven and could accommodate up to four passengers, although there were also two-passenger versions that became known as Bachelor Broughams. This was a vehicle for well-heeled sophisticates, and the original prototype, built by the coach-makers Robinson and Cook of London, was widely copied for men-about-town around the world. Efficiency was achieved by light construction, while comfort in these more advanced carriages was improved dramatically by a single revolutionary development, the invention in 1804 of elliptical steel springs by the Londoner Obadiah Elliot. Elliptical springs enabled carriage-makers to construct stronger, safer and lighter vehicles that provided their passengers with a significantly smoother and faster ride over the new all-weather hard-surfaced macadam roads.

Yet even the largest and most elegant carriages were cramped in relation to mid-nineteenth-century fashions. In closed carriages, gentlemen had the problem of insufficient headroom for their top hats, while fashionable women dressed in crinolines occupied an enormous volume of space, as shown in James Hayllar's painting *Going to Court* (1863), in which the white snowdrift of gowns of two débutantes, travelling in a closed coach, fill the vehicle's interior. Entrance and egress were also made difficult for both sexes by the door openings, which were limited in size to maintain the structural integrity of the body.[3]

Yet it was the open carriage that provided its occupants with the best opportunities to be seen and admired. In 1859 Charles Baudelaire described a typical equestrian scene in the Bois de Boulogne, as painted by Constantin Guys, whom the writer called 'The Painter of Modern Life':

The lightness and flexibility of steel were exploited elegantly in the elliptical springs that made nineteenth-century carriages safer and smoother riding than all earlier types.

The carriage drives off at a brisk trot along a pathway zebra'd with light and shade, carrying its freight of beauties couched as though in a gondola, lying back idly, only half listening to the gallantries which are being whispered in their ears, and lazily giving themselves up to the gentle breeze of the drive . . . Fur or muslin lap around their chins, billowing in waves over the carriage doors. Their servants are stiff and erect, motionless and all alike.[4]

To Baudelaire, the carriage itself was a potent sign of nineteenth-century modernity and of the socio-economic structures within which it was defined. The very different accommodations it offered to passengers and coachmen showed precisely the class relationships in which the modernity of his time functioned. He described how the privileged, relaxed, semi-recumbent postures of the 'beauties' seated in the softly upholstered, boat-like carriage contrasted sharply with the erect, dutiful postures of the coachmen sitting on a hard plank, high up behind the horses. Yet the status of both masters and servants is elevated by the machine above the common mass on foot, while the carriage parties are distinguished from the equestrians around them by virtue of the enclosing bodywork, which also provides them with a stage from which they can communicate, by look, gesture or word, with the other characters on the drive or field.

Although they may appear quaint to the twenty-first-century eye, in their time these were vehicles of the most advanced design. Accordingly, Baudelaire drew attention to the depiction of motion and speed in Guys' drawings and prints, ideas central to the writer's definition of what it is to be modern: 'the fugitive, fleeting beauty of present day life, the distinguishing character of that quality which . . . we have called modernity'.[5] Unlike the present British monarch, moving at walking speed in her antique golden coach, Guys' beauties were travelling fast in the latest machines of their day.

## Horse Power to the People

Before the railway era, ordinary travellers had the opportunity for the first time to enjoy protection from the elements and a semblance of the comfort of coach travel that had previously been available only to nobility. Britain, in particular, made significant advances in the development of stage coaches because of the early consolidation of the Royal Mail, which began to carry post by coach in 1784. More refined vehicles, regular change of horses and the construction of macadam roads in the early nineteenth century all contributed to the increasing efficiency of British road travel and to the relative comfort of passengers who could be carried along with the mail.[6]

Despite these advances, the experience for public-road passengers in any country, before the advent of the car, was hard. Nineteenth-century coach bodies were sturdy, given the technological limitations of wood-and-metal construction, and many were well finished inside and out. Yet their suspension systems were basic; their seats were thinly padded; windows or curtains did not keep out the weather; and the dimensions of the typical interior were cramped. Technical considerations of safety were consistent with the casual attitudes of the time towards all safety issues.

Only in cities, where main roads were paved, were both public and private vehicles able to offer their occupants a reasonable level of comfort more consistently. London's main roads had been paved almost exclusively with granite setts, or cobbles, until about 1840, at which time macadam-cemented gravel roads began to be laid. Although granite was preferred for its durability, it was extremely noisy. Macadam, on the contrary, was soft and quiet, making road journeys considerably more comfortable.

In the second quarter of the nineteenth century the first omnibuses appeared on city streets, in Paris, New York and in rapidly expanding London. The word, 'omnibus' has come to mean 'something for everybody', and indeed the omnibus provided a radically modern environment for the development of a new type of urban culture in which personal interaction became transient, observation casual, and communication impersonal. In the 1830s Charles Dickens commented on the appearance of the omnibus, the 'gaudiness of its exterior, the perfect simplicity of its interior', a pairing of qualities that captures the dualistic nature of the vehicle: workhorse and magic carpet.[7]

The omnibus developed its form slowly, but was spurred forward when in 1856 the London General Omnibus Company (LGOC), the newly formed offshoot of the French Companie Générale des Omnibus, offered a prize of £100 for the design of an improved vehicle.[8] This Anglo-French relationship foreshadowed today's international market for the production and consumption of transport vehicles. The result was a bus of more generous proportions internally, with a full 6-foot standing headroom, yet retaining external dimensions suitable to the many narrow streets of nineteenth-century London and Paris. The interior width was increased from approximately 4 feet 6 inches to 5 feet (137–150 centimetres), allowing more legroom for the centre-facing passengers, whose comfort was also enhanced by the introduction of fixed side windows with vents above them, reducing drafts. On the longitudinal bench seats, passengers were separated in groups of three by a curving brass rail, providing some order to the disposition of bodies. For those climbing up to the seats on the roof, better handles and steps were provided, improving safety and ease of entry and exit. Straw, which was used to cover the floors of early omnibuses, was replaced first by rush matting and then in the improved omnibus by grooved timber flooring, which allowed water to drain away.

The prototypical nature of the omnibus interior is evident in *The Bayswater Omnibus*, painted in 1895 by George William Joy, a view across the aisle of the vehicle and out through the side windows, where a hansom cab can be seen passing at speed beside the bus. Here we see an interior, which, apart from the costumes of the passengers, appears remarkably modern and familiar to the twenty-first-century eye. The interior space is tubular; passengers sit side by side in a row, facing their fellow passengers across a generous aisle, with grooved wooden flooring in the clear space between them. Standing passengers steady themselves by grasping a rail attached to the arched ceiling, while a continuous ribbon of windows provides relief from the crowding and offers a panoramic view of the passing scene. Advertisements are pasted on the wall between the windows and the ceiling.[9]

The greatest competitors of the omnibus were the new underground railways, the first opening in London in 1863, and the horse-drawn rail tram, developed initially in American cities. Trams benefited from the low mechanical resistance of their wheels running over smooth iron rails. The result was that two horses could haul a far larger vehicle capable of carrying twice the number of passengers (as many as 50) of the typical omnibus. Iron rails also offered a more comfortable ride for the occupants, and the small wheels of the tram allowed for a lower floor, easing entry and exit, and making possible a wider, more commodious interior.

By the middle of the nineteenth century urban public transportation networks were established around the world and employed various types of vehicle, developing some local variations in design along the way. In San Francisco, where the city's hilly terrain required the use of cable traction, the car was partially open-sided, enabling passengers to alight directly from their seat to the road. The much-loved San Francisco cable car was an invention of the 1870s, exported in the 1880s to other hilly cities around the world by its patent-holder, Andrew Smith Hallidie.[10] The unique physical sensation of riding in a trolley, of any nationality, is described in the naively erotic lyrics of the popular 'Trolley Song':

Clang, clang, clang went the trolley
Ding, ding, ding went the bell
Zing, zing, zing went my heartstrings
From the moment I saw him I fell

Chug, chug, chug went the motor
Bump, bump, bump went the
    brake
Thump, thump, thump went my
    heartstrings
When he smiled I could feel the
    car shake[11]

The London tram ultimately developed into a very sophisticated and comfortable vehicle. The Feltham model, introduced in 1931, was a double-decker, larger than a contemporary bus. It was well heated and lighted inside and offered a commodious interior with contoured leatherette seats, the backs of which could be reversed to alter the seating plan, enabling all passengers to face the front in rows or to sit face-to-face if they were travelling in groups. The Feltham's handsome interior employed durable but rich-appearing materials in a palette of well-coordinated brown, red and blue combined with varnished woodwork. The driver's station, located behind a curved, panoramic window at the front entrance, was elaborately decorated with curvilinear metalwork. Since this was a prime mover for urban and suburban people, it was intended not just for workday commuting but also for pleasure outings with family, friends and lovers, and its comfortable interior reflected these latter purposes.

This hansom cab, which was used by the architect Sir Charles Barry, boasted fully glazed passenger doors. Such vehicles provided the fastest and most comfortable method of getting around town. The chummy interior seated two in privacy yet offered fine views out to the front and sides. The driver, seated aloft at the back, had an excellent view of the traffic ahead.

Whether drawn by a cable, horse or electricity, streetcars became a beloved feature of city life lasting well into the twentieth century, when increased traffic and the competition of the motorbus made their continued use unsustainable in most places. They were gradually replaced, initially by rubber-tyre trolleybuses, which were capable of pulling up to the curb, rather than depositing their passengers in the middle of the road. Although these survive in some cities, their importance as mass transporters had ended by about 1950.

Urban travellers requiring greater privacy and speed had the option, from the mid-nineteenth century, of hailing a cab. Among the first was a type of four-wheel cab, commonly known as a growler, which could accommodate parties of up to four. But for a more intimate and faster ride, the elegant hansom cab became the vehicle of choice. The hansom or 'safety' cab had been patented in 1836 by the Leicestershire architect Joseph A. Hansom, but was significantly improved by John Chapman and later by F. Forder. In its mature form the hansom was a light single-horse vehicle in which the passengers rode low, between two large wheels, with the driver perched high at the back. Driver and passenger communicated by means of a small trapdoor in the roof of the vehicle.

Entered from the front through a pair of half-doors (or half-glazed full-height doors in some models), and with its side windows positioned just over the wheels, the well-upholstered hansom offered its passengers a 180-degree view of the passing scene. The hansom became not only an icon of nineteenth-century London, but also one of the principal symbols of Arthur Conan Doyle's fictional detective, Sherlock Holmes, since it offered speed and the opportunity to observe the street from its discreet interior. Yet it was manufactured and used around the world, becoming as much a fixture in New York as it was in London.

## The Private Carriage

In the nineteenth century most town dwellers used public transportation as the most convenient way to get from A to B. The narrator of Booth Tarkington's *The Magnificent Ambersons* describes the 'accommodating' nature of the horse-drawn streetcar in North America:

> a lady could whistle to it from an upstairs window, and the car would halt at once and wait for her while she shut the window, put on her hat and cloak, went downstairs, found an umbrella, told the 'girl' what to have for dinner, and came forth from the house.[12]

By the end of the century, however, the tempo of life was speeding up and public transportation was becoming more systematic. Transportation companies imposed timetables and official alighting points for their vehicles. Since individuals had less control of how they used public conveyances in their towns, they became increasingly attracted to forms of transportation that allowed them to do as they pleased. The safety bicycle, patented by John Kemp Starley in 1888, became the preferred transport of hundreds of thousands of sporting individuals. During the second half of the century, particularly in America, the private carriage became more widely available to middle-class families and professional people, the 'carriage trade'. This was facilitated by changes in American manufacturing practices, and the consequent reduction of cost, pioneered by companies such as the Studebaker wagon works of South Bend, Indiana.

Founded in 1852, the Studebaker works became the largest wagon-building factory in the world, producing more than 75,000 units per year by 1885.[13] Studebaker's success was achieved by building its vehicles, including carts, buggies, carriages and wagons, according to the so-called American System of production, based on division of labour and the use of standardized, inter-

changeable parts.[14] Within this system, however, specialist craftsmen carried out specific aspects of construction, including interior wood-work, metalwork, paint finishing and upholstery.

Fifty years before Henry Ford 'put America on wheels' in the early twentieth century, Studebaker and many smaller firms made private vehicles for the middle classes in such great numbers that it became possible for the first time to speak realistically of the owner-driven family carriage, used privately for business and pleasure. The songwriters Rodgers and Hammerstein celebrated the glamour of such vehicles in a song from the popular musical *Oklahoma* (1943):

The classic American surrey featured two well-upholstered sofas under a fringed canopy, an arrangement that defined the interior layout of all later family vehicles. According to the lyricist Oscar Hammerstein, the surrey boasted 'isinglass curtains y' can roll right down . . . '. Such vehicles set a standard for their time in terms of comfort and elegance on a domestic scale.

> Watch that fringe and see how it flutters
> When I drive them high steppin' strutters.
> Nosey pokes'll peek thru' their shutters and their eyes will pop!
> The wheels are yeller, the upholstery's brown,
> The dashboard's genuine leather,
> With isinglass curtains y' can roll right down,
> In case there's a change in the weather.
> Two bright sidelights winkin' and blinkin',
> Ain't no finer rig, I'm a-thinkin'
> You c'n keep your rig if you're thinkin' 'at I'd keer to swop
> Fer that shiny little surrey with the fringe on the top![15]

Thus, it was in Studebakers that the family outing became a national pastime, younger members of the family learnt to drive, and the household shopping was driven home. Tarkington described the popular form of display enjoyed by carriage owners in the second half of the nineteenth century:

> everybody knew everybody else's family horse-and-carriage, could identify such a silhouette half a mile down the street, and thereby was sure who was going to market, or to a reception, or coming

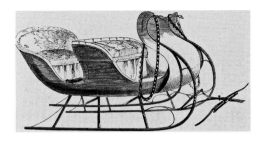

Jingling bells, fur lap robes, colourful tufted and buttoned upholstery and lashings of decorative trimming contributed to the image of the sleigh as a joy-ride vehicle, in addition to the considerable practical advantages of such conveyances in the period before the widespread construction of surfaced roads that could be cleared of snow. This Canadian sleigh is built along English lines and was displayed at the Great Exhibition held in London in 1851.

home from office or store to noon dinner or evening supper.[16]

Studebaker's wide-ranging output also included military vehicles, built in large numbers under contract to the Union Army during the Civil War, Conestoga wagons that carried settlers' households westwards, utilitarian freight wagons and light buggies, which sold from as little as $25 each. They also built carriages for the luxury market, including a low-slung open landau for President Abraham Lincoln and a convertible landau for Lincoln's successor, Ulysses S. Grant. Grant's was blue with yellow wheels and blue silk upholstery to complement the bodywork. Careful coordination of interior and exterior colour schemes was a hallmark of fine carriage design in the Victorian period and set a standard to which automobile makers later aspired.

Carriages took many forms, which did not always involve the use of wheels. In North America, where snowy winters are the norm, the sleigh was a popular type of conveyance. The catalogue of the Great Exhibition held in London in 1851 shows 'a sleigh of elegant proportions' manufactured by J. J. Saurin of Quebec. The catalogue entry attempts to put this exotic vehicle into its regional context:

'Sleighing', as it is termed, forms one of the principal amusements of the Canadians of all ranks who can afford to keep one of any description, and the wealthier part of the community exhibit no little taste, and spare no expense, to put their carriage and all its appointments into suitable condition. The harness of the horse is generally very gay, and beautifully ornamented; while the fur robes in which the riders envelope themselves to exclude as much as possible the severity of the cold, are often very costly . . . The rides and drives around Quebec, Montreal, Toronto &c., are, during the winter months, quite lively with the showy equipages, and musical with the bells suspended from the heads of the horses.[17]

Thus, the concept of travel for pleasure in a beautiful vehicle was firmly established as a right of the middle classes by mid-century. The Great Exhibition catalogue shows several types of owner-driven vehicles, among which one particular type, the phaeton, is shown in three different versions manufactured in Britain and North America. The phaeton was originally devised as a single-horse, two-passenger sporting machine, with a collapsible top; it was the small, lightweight, high-performance roadster of its day. The British examples are extremely elegant curvilinear designs, their bodywork pared down to absolute essentials. They are described as being colourfully painted, with interiors 'lined' in silk, velvet and lace. They all show a 'dicky' seat, appended precariously to the back, to accommodate a groom or an extra passenger. By contrast, the American example, built in Philadelphia, is shorter and boxier, the two passengers perched on a bench with little back support and the thinnest of upholstered cushions. Its upright angularity is typical of the mass-manufactured American carriages of the second half of the nineteenth century. Yet, however different they look, these vehicles were all designed and constructed to offer their drivers and passengers a thrill, an opportunity to demonstrate driving skill and to indulge in the excitement of speed, whether in the park or on the open road.

## Travel for All: The Railway Era

In Britain in 1825, the Stockton & Darlington Railway inaugurated the world's first 25 miles of public rail track, carrying its first fare-paying passengers (in 1829) at a speed of 30 miles per hour. For the rest of the nineteenth century Britain led in the design and construction of locomotives and in laying down track and other infrastructure. Nevertheless, France also developed an extensive rail network due to substantial government investment and central control of the track and signalling network, which was then leased to train operating companies. At the same time in the United States, a railroad-building mania grew as the result of a fervent desire to unify the North American continent, and from the less savoury competitive greed of iron-and-steel magnates and railroad entrepreneurs, who in the process became America's first millionaires. In 1840 the USA had built 2,816 miles of track to Britain's 1,331 miles. Such figures began to increase exponentially as the technology of building rail bridges, tunnels, stations and traffic-management systems developed. By 1875, the fiftieth anniversary of the first commercial rail service, there were 160,000 miles of track in operation around the world.

The network of main and branch rail lines covering the land between cities and towns during the nineteenth century had a profound effect on the economic as well as the physical geography of the communities it served. Advanced engineering works such as Brunel's rail bridge at Maidenhead (1838) and the Forth Bridge at Edinburgh marked those cities as modern centres of commerce and became symbols of England's industrial might. In America, the Central Pacific and Union Pacific railroads met at Promontory Summit, Utah, in 1869, joining the vast North American continent into a single market, if not a single community. It provided the spine to which branch rail lines, stagecoach routes, canals and rivers would feed travellers and goods from north and south. This huge engineering accomplishment was one of the first major portents of the industrial power the US would become in the twentieth century. The transcontinental railroad was the single great link between the east and west coasts until the opening of the Panama Canal in 1914.

The design of railway carriages began conservatively. In Europe, early first-class carriages were designed to imitate the private coaches familiar to wealthy travellers. Three or more coach bodies were normally grafted together on a single four-wheel flatbed wagon, each compartment entered separately. As in the road carriages of the time, their interiors offered a high standard of upholstery and detailing, but their ride over iron rails was considerably smoother than any country roads could offer at the time. Sitting typically three abreast, travellers facing forward or backward could all enjoy excellent views. Augustus Leopold Egg's painting *The Travelling Companions* (1830) shows the interior of a sumptuously upholstered railway compartment occupied by two well-to-do and blasé young sisters sitting opposite one another and flanking the three-part carriage window, which frames the view of a Mediterranean coastal landscape like a triptych. One sleeps, the other reads, neither taking the slightest interest in the passing scenery.

Not all early train carriages, however, provided such panoramas of the landscapes through which they passed. Charles Dickens set down his impressions of a train in which he travelled in America in the 1840s: 'The cars are like shabby omnibuses, but larger' and 'there is a great deal of wall'.[18] His acerbic remarks draw attention to the cattle-car design of early

Charles Dickens observed that the box-like American railroad cars of the 1840s featured 'a great deal of wall'. Spartan amenities for all travellers in the early nineteenth-century American republic demonstrated a commitment to classlessness that would not last long.

American railway carriages, which were large wooden boxes on wheels, their single-class interiors furnished with sparsely upholstered benches equipped with reversible backs, allowing the passengers always to face the direction of travel, but hardly letting them observe what appeared outside. Men and women rode in separate but identical carriages, one filled with smoke, the other not. Both, however, lacked light and views due to their small and widely spaced windows, not aligned with the seats.

At the same time in Europe, although there were plenty of views

Early European trains provided very different levels of accommodation for first-, second- and third-class passengers. The latter travelled in open cars and were offered goggles for sale to protect their eyes from wind and soot. This is Hervé Danmier's cartoon depicting the inaugural journey of the first line from Paris to Orléans on 2 May 1843.

available from the open carriages in which they rode, second- and third-class passengers and the train's crew fared even less well in terms of comfort and safety. Second-class accommodation in the 1820s and '30s generally offered wooden benches and a roof overhead in a car otherwise open to the elements and to the emissions of the locomotive. Third-class passengers rode in open wooden carts in which they had to stand. Honoré Daumier's engraving of 1843, *A Pleasure Trip from Paris to Orléans*, lampoons the opening of one of France's first inter-city railway lines by detailing the harsh experience of passengers riding in open carriages and attempting to protect themselves from the irritating effects of the weather, wind and smoke. They huddle forward, heads bowed, top hats held tightly against the force of the wind, faces shielded by scarves and eyeglasses. Yet from the 1850s many thousands of people were lured on board trains for the first time to visit a series of spectacular international fairs, such as the Great Exhibition of 1851, which attracted many people from the north of England and abroad to travel to London.

The train interior soon became a meaningful setting for narrative paintings such as Abraham Solomon's companion pictures, *First Class – The Meeting* and *Second Class – The Parting*, which illustrated the differing amenities available to rail passengers of upper and lower classes. Solomon's first-class compartment is distinguished mainly by its upholstery, which is generously padded and tufted and finished in a luxurious patterned fabric. The seats feature padded armrests and winged head supports, separating each passenger's personal space. Window frames

Abraham Solomon's paintings *First Class – The Meeting* and *Second Class – The Parting* (1855, Southampton City Art Gallery) illustrate the differences in travel accommodation enjoyed by the rich and poor in the second half of the nineteenth century in Europe. As usual, the rich are shown to be happy and flirtatious, while the poor are sad and anxious. In the first-class compartment the trappings of domesticity disguise the mechanistic nature of the train. The poor, by contrast, experience the harsh reality of the machine.

are elegantly curved with tasselled cords to open or close the curtained panes. Here Solomon portrays a romantic atmosphere of chocolate-box sweetness. By contrast, the second-class compartment is austere, its windows unshaded and benches of bare wood, its only decoration the advertising posters, aimed at emigrants to Australia. The mood here is sombre and anxious, the characters' emotions set off starkly by the hardness of the interior.

The anonymity of modern travel led, inevitably, to various forms of moral lapse. The train *de luxe*, with its bordello decor, illustrated in a French magazine around 1900, was portrayed as a perfectly constructed environment for sexual dalliance. A well-dressed young woman catches the eye of a dashing fellow traveller across the aisle while her elderly husband concentrates on the stock-market report in his broadsheet newspaper. Here was a place designed to encourage conversation or even flirtation with strangers, a place where time passed slowly but the everchanging scenery reinforced the sense of suspension from everyday concerns and the constraints of home. As Freudian interpretation later testified, the speed of the train, its gentle rocking and the rhythmical rumble of the wheels contributed physically to the erotic potential of the *wagon-lit*.

Over a fifty-year period travellers came to terms with a threefold increase in the speed at which they were able to move. The average rail speed in the early 1830s was not much above 15 mph, but by 1890 British trains were capable of 75 mph and achieved average journey speeds of between 50 and 60 mph. Ordinary people, riding at such speeds and viewing the landscape through the side windows of their carriages, experienced the world in an entirely new way.

The train window framed broad vistas of landscape seen as a slowly changing panorama while the trackside whizzed past in a blur. Signal boxes and trees near the line all took on an abstract appearance previously unknown in the world of common experience. Repeated elements, such as telegraph poles at the side of the track, established visual rhythms and sound repetitions that complemented the mechanical *clickety clack*, emitted by the wheels rolling over joints in the iron rails, and the continuous *chuffing* sound of the steam engine. Trains passing in the opposite direction presented an even more dramatic spectacle, hurtling by in a rush of fragmented forms, colours and lights accompanied by a Doppler-effect crescendo of wind and mechanical noise. Thus the traveller's gaze was changing fundamentally towards a proto-cinematic experience for an immobilized armchair spectator.

## The Pullman Environment

Until the advent of the automobile, there was no real competition with the railways for the movement of cross-country passengers. Although technological development was relatively rapid, the train interior took more than forty years to develop a level of comfort that we might recognize as acceptable by today's standards. Meanwhile, the railway carriage developed in two directions. The industrial designer Corin Hughes-Stanton, writing in the 1960s, classified train interiors in two types: (1) transporters, vehicles designed to accommodate the maximum density of seated passengers in spaces strictly limited by technical and economic factors; and (2) 'places which move', vehicles that provide the characteristics of rooms in a building, such as lounges, bedrooms or dining rooms.[19] Railway carriages came to encompass both types.

The first and perhaps the best model of trains that offered their passengers the experience of a 'place' was the series of royal trains built for Queen Victoria, who was among the first European monarchs to use a train as a practical means of getting around. Yet her sense of what was 'practical' differed from ours today. The first royal saloon carriage, built by the Great Western Railway for the British royal family in 1842, was decorated in the style of Louis xiv with a colour scheme of crimson and white silk. Each compartment had a specific function – bedroom, reception, staff accommodation – allowing the royals' life and work to proceed while they were on the move, just as they would do at home in a palace.

The decor of a nine-compartment interior in a double carriage used by Victoria during the 1870s crosses the funeral parlour with the brothel. The queen's day saloon was completely upholstered with tufted satin over the coved ceiling, walls, sofa and chairs. All this sumptuous upholstery was combined with innovative construction – the floors were a sandwich of timber and cork – to reduce noise and vibration. Victorian technology was also applied to the comfort of the ladies-in-waiting, who were provided with convertible sleeper seats used on longer journeys. Tins filled with acetate of soda crystals, reacting chemically with boiling water, provided heat.

Windows were elaborately curtained, lampshades draped, cushions scattered, cords tasselled. The textiles were mainly silk and satin in various shades of the favourite royal blues and gold, while the timber furnishings were in Bird's Eye Maple. Gilded oil lamps, soon converted to electricity, provided artificial lighting, while broad windows ensured a good view both in and out of the carriage. By attempting to recreate the atmosphere of a reception room in a royal residence, with all its pomp and formality, in the smaller, longitudinal space of the carriage, train interiors became an entirely new type of architectural 'place'.

Across Europe, the industrialized monarchies of the second half of the nineteenth century competed to produce the most elaborate and impressive rail carriages, just as they competed in other matters of taste and industry. Even Pope Pius IX was involved in the game. The mobile throne room, built for him by the Pio Latino Railroad, included a domed Baroque *baldacchino* over a white velvet travelling throne. This was the first motorized Popemobile.

According to Sigfried Giedion, George Mortimer Pullman may have been thinking of the contemporary *train impérial*, built for Napoleon III by the Compagnie de Chemin de Fer de l'Est, when he designed his first palace cars in 1859. The *train impérial* had been widely published two years earlier in an elaborate folio with colour illustrations, and it is likely that Pullman would have seen this. Napoleon's train was constructed by one of the greatest French engineers of the time, Camille Polonceau, and decorated inside and out by the archi-

tect Eugène-Emmanuel Viollet-le-Duc, General Inspector of Historic Buildings and the ultimate authority on French taste and decoration in the second half of the century.[20]

The carriages of this train were higher and wider than the standard at that time, their increased dimensions offering Viollet-le-Duc the opportunity to create more palatial interiors than would be possible within the typical dimensions of contemporary trains. He employed his extensive knowledge of botany and zoology to inform the decorative scheme for the furnishings, metalwork and other elements in creating a harmonious set of interiors, just as he would have done for the design of rooms in a grand château. The elaborately articulated ceiling decorations and large-scale furnishings of these carriages produced the impression of a series of 'places', linked to each other by the latest American-style open verandas, enabling the royal retinue to move from one car to another while the train was in motion.

The French emperor Napoleon III enjoyed an even grander travelling experience than Queen Victoria in his *train impérial*, decorated by Eugène-Emmanuel Viollet-le-Duc in 1857. Indeed, this was the most elaborate rolling palace of its day and set a standard of luxury accommodation that would profoundly influence the development of commercial railroad cars over the following half century.

In the dining car, which included an office and a washroom, the furnishings were in carved oak, the embossed and gilded leather-covered walls draped with silk, and the ceiling painted with clinging ivy. Next door, the observation car was an open, flatbed carriage, its roof supported by polished iron columns linked by cast-iron balustrades decorated with gilded foliage. Tapestry curtains, normally tied back to offer the best views of the passing landscape, could be closed to shelter occupants from the wind or to provide privacy. A *salon d'honneur* carried an armoury of decorative weapons and was surmounted by a gilded crown supported by eagles. Furniture was in carved mahogany and walls were lined with green silk damask. The ceiling, springing from a cornice in carved and gilded oak and supported by gilded bronze pillars, was decorated with the idealized figure of the emperor, entwined in the gilded branches of a bay tree.

Whereas the dining car and *salon d'honneur* were single-purpose spaces, Napoleon's sleeping car demonstrated the potential of convertible furniture for the efficient use of space in trains. A cutaway drawing, showing the bedroom of the emperor and empress, reveals a pair of carved ebony beds, which were treated as miniature rooms within the larger space of the carriage. They were curtained with sky-blue and garnet-red velvet and could be separated from the larger space by panelled, folding doors. Between them was an anteroom leading to individual bathrooms. Another suggestion of things to come, the accommodation for the ladies-in-waiting was furnished with seats that could extend and fold to form a bed, like the military campaign furniture that became fashionable during the early nineteenth century.

The technical development of comfort in rail travel began with rapid advances in patent furniture during the mid-nineteenth century. The study of ergonomics led to the design of many types of adjustable and folding chairs adapted to a variety of purposes, from office seating to the barber's chair. When applied to the traveller's seat, the principles of ergonomics demanded the utilization of several features that became standard. These included a reclining seatback, padding for lumbar support, contoured headrests, swivelling and tilting seat mechanisms, armrests and movable footrests. During the 1850s in America, engineers and inventors employed all these features to devise a railroad chair that would support the body comfortably in a variety of positions for reading, sleeping and conversing during long journeys. Patents were so many and so varied – and some so bizarre – that Giedion described them as 'the Jules Verniads of furniture'.[21]

The folding bed, too, had been a preoccupation of inventors for decades before it was applied to railroad comfort. Multi-purpose furniture had been marketed as practical and economical for use in hotels, smaller apartments, boarding schools and other places of temporary accommodation, as well as in canal boats and riverboats. In a spirit of efficiency, invention and imagination, folding beds were combined with or disguised as wardrobes, dining tables and even pianos.[22] More sensibly, Theodore T. Woodruff devised, patented and built the first practical convertible railway furniture for a series of sleeping cars, which he operated successfully in the 1850s. But it was Pullman who designed the convertible sleeping car that became the industry standard. In 1865 he constructed an experimental luxury 'Palace' car, *Pioneer*, which broke the mould of earlier commercial railway carriages.

Like the *train impérial, Pioneer* was higher and wider than anything else on rails. Its distinctively shaped roof provided a longitudinal clerestory illuminating the central aisle. In the daytime, passengers were seated in pairs of couches facing each other. These were arranged on both sides of the aisle. Over each pair of couches, an upper berth was hinged above the window and fastened to the roof at a 45-degree angle, its underside panelled and decorated to form part of the ceiling. At night the car converted from parlour to dormitory by folding the couch seats to form a double berth, by lowering the upper berth to a horizontal position, and then by drawing curtains around each berth for privacy. *Pioneer* became instantly famous on its inaugural run when it carried the body of the slain President Abraham Lincoln from Chicago to his Illinois home town for burial. Photographs and drawings of the car decorated as a hearse, draped in black satin, were widely published and immediately established its prestige.

Following the success of the convertible sleeping car, the comfort of the Pullman train was enhanced in the 1880s by a further burst of technological developments. Full electric lighting and through-train steam heat were introduced to create a completely controllable environment. In 1887 Pullman patented the enclosed, concertina-style vestibule connecting all the carriages of the train. This seemingly simple device joined the individual cars into a flexible, monolithic structure and a unified linear space, through which passengers could move for the first time in comfort and safety.

The concertina vestibule paved the way for Pullman to introduce a range of single-purpose cars including club car, restaurant and observation car, and additional facilities such as library and hairdressing salon, 'places' that provided service, diversion or conviviality. This is illustrated in a Union Pacific Railroad advertisement of 1879, titled 'Life in a Palace Car', which depicts a group of people, including fashionably dressed ladies and gentlemen, a child and an Indian chieftain, gathered around an

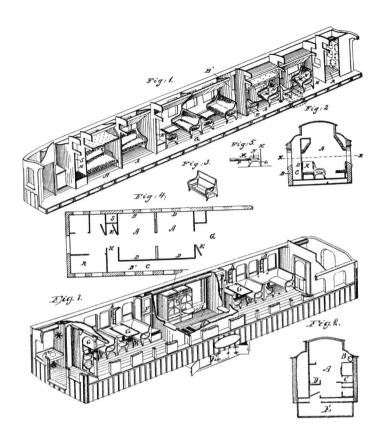

Invention and innovation were the hallmarks of George Pullman's first 'Palace' cars of the 1860s. His patent drawings show how the comfort of the *train impérial* was made available, through the application of new technologies within a commercial system of production and consumption, to any traveller who could pay a modest surcharge to the normal fare.

elegant piano and singing from songbooks. More than any other single entrepreneur or inventor, Pullman defined the modern notion of luxury travel. His aim was to make, as much as was technically possible within the envelope of the train, an environment resembling a good hotel or the first-class section of an ocean liner.

By the 1880s a coach passenger seated on a long-distance train, either in Europe or North America, would often be travelling at a mile a minute, relaxing in an adjustable, reclining railway chair with an elevated panel supporting the feet, viewing the rapidly passing landscape and enjoying an experience unavailable by any other means at the time.[23]

By the 1870s all the major American railroads operated trains that were made up entirely of Pullman cars with private sleeping compartments and, hence, designated 'Limited'. Railroads competed on these services to provide their passengers with the best food and drink available on wheels. Famous restaurants, such as Delmonico's of New York and the chain of eateries opened by Fred Harvey, Harvey House, along the

Chicago to Los Angeles route of the Santa Fe Railroad, provided chefs and menus for great trains, such as the Santa Fe Railroad's *Super Chief*, the New York Central's *Twentieth Century Limited* and the Pennsylvania Railroad's *Broadway Limited*. The dining-car interior was typically the most festively decorated car of a train, its tables arranged beside the windows and set with crisp linen, sparkling china, silver and crystal.

During the heyday of railroading, train interiors both led and followed fashions in architecture and furniture design. Pullman had been a cabinet-maker by trade, and his designs for the early Palace cars of the 1860s and '70s were highly crafted in the Victorian taste. *Pioneer* was ornately panelled in black walnut and featured cut-glass chandeliers and marble hand basins. Pullman's taste for ornament, which, according to Hughes-Stanton, 'stresses the pleasure aspect of travel', started a fashion that by the 1890s resulted in elaborate decors rivalling the European royal trains of the time. Yet not all critics were admirers of Pullman's heavy Victorian ornament. Giedion wrote:

> Like the fashionable [architectural] interiors of the period, the sleeping cars and dining cars succumbed in the 'nineties to superficial luxury. They were overfestooned with Rococo ornamentation, and their high curved ceiling – known as 'full empire' – masqueraded as a stone vault.[24]

Giedion was echoing earlier comments by Edward Bok, one of the most influential taste-makers in America at the turn of the twentieth century through his role as editor of the *Ladies' Home Journal*. Bok condemned the 'veritable riot of the worst, conceivable ideas' in the new trains and was particularly red-faced because the interiors of these trains not only echoed the prevailing tastes of the time, they set the tastes. He commented that nouveau riche travellers were imitating in their homes the types of decoration that they had seen when travelling, just as they do today. 'Every foot of wood-panelling was carved and ornamented . . . gilt was recklessly laid on everywhere . . . Mirrors with bronzed and red-plushed frames were the order of the day'.[25]

Yet critiques based on connoisseurship missed the point of popular luxury, whether in railroad trains, steamboats, theatres or saloons.

An American office desk chair, dentist's chair, barber's chair, operating table, wardrobe beds and various other folding or convertible furniture types, patented in the nineteenth century, provided the basis for a comfortable railroad traveller's chair, which in turn influenced the design of adjustable aeroplane and car seats in the twentieth century.

Despite the advanced technology of his 'Palace' cars, Pullman's background as a cabinetmaker led him to adopt and promote an elaborate decorative style for his carriage interiors. Pullman cars not only followed architectural styles of the past, but presented a new image of eclecticism and sumptuousness that deeply influenced domestic taste up to the First World War.

According to Russell Lynes, writing in the 1940s, these interiors were not works of high-minded aesthetic design, such as true architecture. Instead, they were more akin to the fairground or the music hall, places of pleasure, joy, sensuality and adventure. He wrote:

When [a man] ventures forth either for business or pleasure he moves into a world where he is wafted on swan boats and bedded down in crystal palaces, where he is entertained by women as beautiful as angels (if not so discreet) to the sounds of erotic music and the tinkling of glasses. For the moment he loses himself in the fairyland of the carnival – a prince whose comfort is the first concern of a retinue of servants, and whose eye is filled with riches by scores of artists. In such surroundings and in such delights what matters it to him whether what he beholds be 'tasteful'.[26]

It was not the aesthetic aspects of nineteenth-century railway carriage interiors that worried others. The work of Pasteur and Lister in the second quarter of the nineteenth century laid the foundation for a theory of disease transmission that focused on dust as one of the main carriers of tuberculosis, cholera and typhus. The heavily upholstered and carpeted, panelled and ornamented compartments of first-class railway carriages had the potential to harbour all the deadliest bacilli. As pointed out by one French health reformer in 1907,

The seats and furnishings of carriages are in cloth or velvet; and as if these materials were not enough to retain dust, they are luxuriously upholstered in such a way as to multiply the corners inaccessible to cleaning done with even the best will. Certainly there would be no comfort in these first class carriages for the well-to-do without, for example, thick carpets; but these only aggravate, by their dangerous filthiness, the general insalubrity of the surroundings. It is enough on

Walter Gropius designed elegant and restrained interiors for this German Mitropa rail car in 1914. Its flush veneered wall and cabinet surfaces were enriched by the natural grain of wood and complemented by bold geometric upholstery patterns and jewel-like metal hardware. The sleek interiors echoed the streamlined locomotive casing also designed by Gropius.

a clear, sunny day to observe what happens in a carriage on the entrance of a passenger, and especially of a lady passenger, to get an idea of the dirtiness of such a carriage, said to be luxurious. Each step of the passenger on the carpet, each movement, whether putting down the suitcase, or sitting down, is the occasion for a cloud of dust to arise from the surface touched . . . And as it must be supposed that convalescents with contagious skins, people with colds, flu, and above all tuberculosis are lodged in these compartments, one can easily deduce the qualities of the air that are going to nourish, for a greater or lesser length of time, the newly arrived passengers.[27]

These comments echo debates raging at the time in all branches of the design world, including architecture, furniture and other products. In the twentieth century critics of various persuasions called for a more rational approach to designing new building types, new furniture types and new product types. In Germany, the Deutscher Werkbund promoted excellence in design for both craft products and manufactured goods. Its leading exponents, such as Peter Behrens and Walter Gropius, applied its principles to their plans for transport vehicles in addition to architecture and other products.

Gropius's design for a sleeping compartment in the Mitropa rail car of 1914 rejected the Pullman aesthetic in favour of a simple design scheme that created a luxurious impression through the fineness of its detailing, the richly grained wood veneers covering its flat wall surfaces and its elegant geometrically patterned upholstery. The Mitropa compartment recalls the character of the purposeful and luxurious Louis Vuitton wardrobe trunks of the period, their compact interiors purely functional, ingenious and immensely seductive due to fine craftsmanship and excellent materials.

Despite the forward-looking experiments of the Deutscher Werkbund and the Bauhaus and the published pleas of modernist architects for a rational, modern design aesthetic, until the 1930s luxury trains continued to rely on dazzling ornamental effects and traditional notions of luxury to lure prospective passengers into paying relatively high fares. In Europe, the Orient Express had been in service since 1883, becoming perhaps the most fabled and romantic train of all time.

The exclusivity of the Orient Express was manifest in its small size. This service normally carried only three or four passenger coaches, accommodating up to 28 passengers – but it accommodated them in a grand manner. In addition to a dining car, with ladies' parlour, smoking room and library, *wagon-lit* cars (the European equivalent of Pullman's sleeping cars) contained private compartments, which converted from drawing rooms in the daytime to bedrooms at night.

In their earliest form, the carriages were decorated elaborately, in a manner similar to the American Pullman cars of the period, featuring seats upholstered in leather, embossed with gold patterns. There was carved wall panelling with painted mural inserts, hanging tapestries and oil paintings to create the atmosphere of a grand hotel or stately home. Dinner was a formal, three-hour affair catered from a *fourgon* car, packed with delicacies. Attendants were always at hand to satisfy the traveller's every requirement. And the travellers, themselves, contributed the festive finery of their costumes to the opulence of the train's decor.

The mythology of the Orient Express was based on the Simplon Orient Express, which ran a Paris to Istanbul service from 1919 to 1939 and whose style was a product of the fashionable taste of those years. Today's popular image of this train was formed mainly by the successful 1974 film version of Agatha Christie's novel, *Murder on the Orient Express*, although the train carriages that appear in the film were not original stock but rather Type LX *wagon-lit* carriages of 1929, as would have been used on any of the major European trains *de luxe* before the Second World War. The popularity of the film stimulated a commercial interest in nostalgic luxury train travel and led to the restoration of similar LX carriages for the newly founded Venice Simplon Orient Express Company (VSOE), which began its current London to Venice service in 1982.

The film's train interiors are decorated in the Art Deco style, with elegant marquetry wall panelling, delicate metal-and-glass fern-patterned light fittings, Lalique-style cast-glass relief panels of nude figures inset between the windows, and colourful, geometrically patterned upholstery. The Art Deco style, particularly of the dining and salon cars in which much of the main dramatic action of the film occurs, is contemporary with the story, set in 1935. On the current Venice Simplon Orient Express, the decor of the refurbished train is in the earlier Edwardian style and includes mosaic flooring, marble basins in the cloakrooms, moiré blinds at the windows, brass lamps with pink silk shades, woven Florentine upholstery, satin sheets and lace pillowcases, a setting contrived for romance and consequently popular for honeymoon and wedding anniversary trips.[28]

## Late Trains

In America, the railroading age peaked between 1930 and 1950 with a generation of dramatically modern streamlined trains, intended to offer their passengers an image of luxury based on high technology rather than historical models or the French Art Deco. The period's great trains, such as the *Twentieth Century Limited*, the *Broadway Limited* and the *Super Chief*, operated in a highly competitive market threatened by the rise in car ownership, by the construction of interstate highways and by the nascent airline industry. This situation continued to worsen for the railroads after the Second World War and culminated with the inauguration of cheaper jet passenger services by all major airlines around 1960.

Yet the culture of railroading had become firmly established across all classes of society and, for a time, maintained its allure through the introduction of trains that captured the spirit of modern travel in the increasingly design-aware climate of Depression America. Beginning in the late 1920s, a group of young commercial artists built the foundations of a new profession, industrial design. Among the most successful of these, Walter Dorwin Teague, Norman Bel Geddes, Raymond Loewy and Henry Dreyfuss brought together large teams of talented individuals with expertise in fashion and theatre design, engineering, packaging, architecture and display to enable their consultancies to address the needs of clients in just about all branches of manufacturing and retail activities as well as the public sector.

In the harsh Depression economy following the Stock Market Crash of 1929, American manufacturers and retailers needed the competitive edge for their products that was now being offered by consultant industrial designers. A strong relationship between modern design and advertising was forged to encourage consumption not only of domestic goods, but also of services such as those provided by the transportation industries.

Railroads quickly engaged the skills of industrial designers to recreate their image, to provide a new look for their locomotives and passenger cars, to generate publicity and to attract travellers. The designers' approaches were similar in one significant way; in the 1930s they all employed the pseudo-science of streamlining as a foundation for their aesthetic. In train design, streamlining, as a means of reducing air resistance and improving performance, can be traced back to a project of 1865 by the Reverend Samuel R. Calthrop, whose extraordinarily prescient patent drawings showed a long tapering locomotive pulling a set of smoothly skinned carriages that bear an uncanny resemblance to the most advanced trains of today (see p. 56).

By the late 1920s wind-tunnel testing of rounded, tapering and teardrop shapes for airships, aircraft, automobiles and trains had become a relatively sophisticated method of refining designs to achieve increased speed and economy of operation. Universities, including the Massachusetts Institute of Technology and the University of Michigan, had established wind-tunnel facilities, which were used by industrial designers to obtain data on which they justified designs that may, in fact, have been based as much on taste as on scientific evidence.

The implications for the interior design of train cars, however, were significant. Streamlined cars had rounded ends, cross-sections that tapered towards the top and domed rooflines that created challenges and opportunities for designers wanting to maximize the actual and apparent space of their interiors and optimize the arrangement of furniture. Designers also exploited the internal curves, imposed by the external forms of the cars, to dramatize the modernity of these spaces. One of the earliest successful American streamliners was the diesel-powered *Burlington Zephyr*, built by the Budd Manufacturing Company in 1934. The thin-walled, lightweight, stainless-steel shell of this train helped to make it both fast and economical and also gave it the appearance of speed, which made it instantly popular when it was first seen by many thousands of visitors at the Chicago World's Fair and as the star of a feature film, *Silver Streak*, in the same year.

Paul Cret designed the interiors of the fully air-conditioned train, which through its wide publicity contributed to the growing awareness of modern furniture and interior design. In an article, published by the *Magazine of Art* in 1937, Cret discussed in some detail the relationship between structure, materials and style in a modern train interior:

> our intent was to use the engineering design of the cars as a basis; the thin metal forms, the streamlined outer envelope, the inner surfaces of thin material frankly a veneer over the structural shapes, the intervening voids being filled with insulating material. The thinness of these veneer materials necessitated the use of countless screws whose heads must be part of the resultant impression, or of cover moulds of the same stainless steel, or of aluminium. In some of the cars, this veneer material – Masonite or Homosote, or where slightly larger dimensions were required (e.g. for ceiling surfaces), aluminium sheets – were [sic] finished with a dull surface enamel paint, sprayed on. In other cases these surfaces were in part covered in Flexwood. As the latter could be had in many veneers, in a wide variety of color and marking, it was possible to have the small rooms of a compartment sleeper, each in a different color scheme, thus doing away in large measure with the monotony of the earlier sleeping cars ... [T]he general impression of the interiors was one of bright color accented with lines of chromium or stainless steel mouldings.[29]

A French-born architect trained in the Beaux-Arts system, Cret became Head of the Architecture Department at the University of Pennsylvania. Although he was sympathetic to the European modernist architects of his generation, Cret was above all an artist who worked in imaginative ways beyond what he saw as the limitations of European Functionalism. In addition to three Zephyr trains for the Burlington Railroad, he designed interiors for the Santa Fe Railroad's flagship, the *Super Chief*, which began service from Chicago to Los Angeles in 1936. Writing about this train, he described the regional influences in the design of the bar car, those elements intended to give the interior character, while also addressing the practical imperatives of all industrial art.

> Here the colors were those of the Pueblo – the brownish red tones taken from the old pottery, with notes of the bright blues, the blacks and the 'bayetta' red found in the old weavings. The ceiling lighting was a 'plumed serpent' from the Hopi tribal snake festival, to be lighted by blue lamps of low voltage. On the walls were painted rep-

The Pennsylvania
Railroad's K4s
locomotive 3768
in the streamlined
cladding designed
by Raymond Loewy
in 1936.

resentations of the ceremonial dolls of the Pueblo tribes. For the
upholstery, two-pile fabrics were designed, of mohair material to
withstand railroad usage, but of color and design inspired by Indian
weavings . . . Though the finished effect is one of colorfulness and
gayety, with a sense of luxury and comfort . . . all the materials and
the general design must be economical in the true meaning of that
term – good value for the money expended.[30]

Cret collaborated with Raymond Loewy on the design of interiors
for one of the two most sensational trains of the streamliner era, which
were launched simultaneously in 1938 in direct competition with one
another for the prestigious New York to Chicago overnight route. Cret
and Loewy designed the Pennsylvania Railroad's *Broadway Limited*,
while Henry Dreyfuss and Associates designed the New York Central's
*Twentieth Century Limited*. These two trains set the ultimate standard of
comfort and style during the remaining great days of passenger rail travel.
Loewy and Dreyfuss created dramatically streamlined shells to cover the
high-powered steam locomotives of these first-class trains, and both
designs have become stylistic icons of their time.

Passenger cars were then designed to complement the locomotives,
while their interiors reflected the most elegant contemporary style of
furnishing and decoration, using up-to-date materials including
chromium-plated tubular steel, clear plate glass, large mirrored surfaces
to amplify the apparent width of the cars, sound-absorbing cork on the
walls, concealed indirect lighting and simple, woven textiles. The over-
arching theme was Streamlined Art Deco, a style that effectively
transcended class distinctions.

The essence of Loewy's approach to the design of transport vehicles and vessels was to express the modern spirit. He tamed the image of speed, conveying its power and reliability through sleek forms, a new ornamental vocabulary and surfaces full of tactile invitations. The lounge and bar car of the *Broadway Limited* demonstrated his design style of the 1930s in its most appropriate context – travel.

Both in plan and cross-section, the car employed the theme of radius curves, which swept around corners and up into the stepped coving of the ceilings, increasing the apparent space by eliminating most angular junctions. The bar itself was a drum shape, its form emphasized by parallel and concentric decorative bands, linked visually to the smaller furniture and decorative elements of the car by the repetition of circular shapes, as well as texture and colour. Built-in furnishings were simple in form and restrained in detail, using materials that were both purposeful and luxurious. Wood veneers in chequerboard patterns covered the walls. Polished wood and leather upholstery were combined with large mirrors, chromium trimming and newly developed Formica, all illuminated by a subtle combination of natural light, entering through slatted Venetian blinds, and indirect artificial light. In this way, Loewy and Cret provided passengers with a sophisticated environment in which to enjoy a cocktail and a cigarette, converse, and watch the landscape glide past.[31]

Loewy wrote of the rearward-facing observation car he designed for the train. This staggered-level, cinema-like space

Raymond Loewy's and Paul Cret's interiors for the *Broadway Limited* employed modern, weight-saving materials to create a sleek, sophisticated environment for transcontinental travellers in the 1930s. Loewy's style took the machine aesthetic and tamed it for popular consumption. He called this approach MAYA (Most Advanced Yet Acceptable): in other words, pushing the boundaries as far as possible within the limits of popular acceptance in a commercial world.

provided a clear view of the tracks fast disappearing in the distance, quite a thrill for riders when first introduced. A glassed-in super-structure equipped with armchairs made it possible to watch the sinuous motions of the long train taking curves at high speed along the beautiful shores of the Hudson River, an unforgettable experience especially when the leaves turned in fall. Many of these cars were built and operated, adding interest and, some said, romance to American railroading.[32]

By the 1950s the age of steam was over. Ninety per cent of all main-line American railroad trains were being drawn by enormously powerful diesel locomotives, while branch-line trains were powered by electricity. The big diesel locomotives were referred to unromantically as 'hood units', 'road-switchers' or 'blocks', with as many as four of these machines coupled together to pull fifteen or more carriages at high speeds over the most challenging terrain, such as the Rocky Mountains. Increased speed was accompanied by greater comfort, luxury and service in the passenger carriages of the 1950s. Compartments were now equipped with welcome innovations such as electric shaver sockets, private showers and air conditioning. Although in 1931 the Baltimore & Ohio Railroad had launched the first fully air-conditioned train in the world, the *Columbian*, an all parlour-car express running between Jersey City and Washington (later extended to Chicago), air-conditioning did not become a standard amenity until the 1950s.

With the added power of diesel locomotives, post-war trains carried a greater number and variety of activity cars, including themed lounges, bars, larger hair salons, cinema, various diners and spectacular 'dome' cars, two-level observation cars with raised glass roofs. From the elegantly furnished upper-level bars and lounges of the dome cars, 360-degree views of the western landscapes, mountains and deserts provided passengers with unforgettable experiences that had to be seen at leisure to be fully appreciated. This was travel devoted in every way to the pleasure of the passenger en route, the only card in the railroads' deck to compete against the speed of flying – and ultimately a losing hand.

The development of style in the American train interior was graphically described in a photo essay, published in *Popular Science* magazine, which compared interiors of the New York Central's *Twentieth Century Limited* in 1902 and 1952. During that half-century a radical change had taken place, not only in the size of passenger trains but also in the popular tastes their interiors reflected. In the Dining Car of 1902, a Pullman-style ceiling and stained-glass windows, illuminating acres of moulded dark wooden panelling, gave way by 1952 to smooth walls, a

coved ceiling with indirect fluorescent lighting and Venetian blinds at the windows. The earlier Lounge Car's brass trimmed balcony at the rear of the train was replaced by a closed observation car with a smoothly curved wall of glass at the rear. The Barber Shop and Office of 1952 looked less like Edwardian gentlemen's clubrooms and more like laboratories. While the earlier train was charming, elegant and intimate for its 27 passengers housed in five cars, the version of 1952 treated its 160 passengers, in twenty cars, to unpretentious comfort and practical styling.[33]

Radical experiments in Europe also contributed to the development of more modern trains. Germany led with projects such as the Rail Zeppelin, designed in 1930 by the naval engineer Franz Kruckenberg, who aimed to fuse railway and aeronautical technologies to devise a state-of-the-art transport vehicle of an entirely new kind. This was a low-slung, propeller-driven rail car of aerodynamic form and lightweight construction using a metal and ash frame and impregnated sailcloth skin, with an aluminium nose and flush detailing on the exterior. Its BMW aircraft engine powered a single large wooden propeller mounted at the rear. Kruckenberg's efforts bore fruit, as demonstrated on a highly publicized speed run in 1931, when the Rail Zeppelin prototype achieved the world record speed of 230 km/h, which remained unbeaten until 1955.

Like contemporary passenger aircraft, the interior of the car was very narrow, a mere 2.5 metres across, leaving space for only a single seat on either side of the aisle and restricting the number of passengers to 24 in two compartments, smoking and non-smoking. Windows were sealed, double-glazed units, fixed shut due to the speed at which the Rail Zeppelin travelled. Ventilation was achieved mechanically through two longitudinal rows of small circular vents in the ceiling, while lighting was arranged in continuous tubular strips fitted just above the ribbon of windows on each side of the car.[34]

The hygienically austere but elegant interior had the look of a Bauhaus classroom, fitted with its two rows of tubular steel chairs, in the style of Marcel Breuer, undecorated white walls and ceiling, dark polished floors, and the car's slender structural frame visible throughout. Despite its spectacular appearance, its speed and the huge public attention it drew, the potentially dangerous propeller-drive doomed this car to the status of a historical curiosity, and it never went into commercial service. However, it quickly provided inspiration in the United States for other radical train designs, such as the *Burlington Zephyr* and another experimental vehicle that also used aircraft technology, William Stout's Pullman Railplane (1933), with a welded tubular steel framework, a streamlined zeppelin-shaped fuselage and a Functionalist-style interior, with tubular steel seating, similar to the interior of Kruckenberg's 'flying train'.[35]

This advertisement for the Western Pacific Railroad cannot exaggerate the astonishing character of the dome cars operated by many North American railroad companies in the 1950s. These cars were miracles of modern materials and construction technology, and they gave their passengers unforgettable, panoramic views of the continent's most dramatic landscapes.

Although they were less exciting, the majority of trains on which ordinary people travelled many millions of miles in the years from the first World War to the jet age provided good standards of performance, safety, comfort and pleasure. The ubiquity of the sleeping car ensured that those making overnight journeys could, for a modest fare, enjoy the comfort of a good night's sleep in a private berth. In the daytime they could sit in an upholstered armchair and view the scenery outside, dine in a well-catered restaurant car, and find easy access to a lavatory and washbasin. There was steam heat and electric lighting, and porter service was available to all travellers.

While passengers were having an increasingly comfortable and enjoyable experience riding the trains of the inter-wars years, the lot of the engine crew improved less quickly. At the turn of the twentieth century, the drivers and crew of main-line locomotives in Britain were still working in the open air, with the benefit of only a small covered area of the cab to protect them from the worst of the elements. The locomotive crew also formed a strict social hierarchy on the footplate (the cab) of the steam locomotive. The driver was in supreme command behind a display of gauges and controls. The stoker was the hard labourer in the middle, while the engine cleaner had one of the dirtiest, lowliest jobs in history. Part of his appalling routine was to climb inside the coal box to empty ash from the spent fuel during maintenance stops.

Their lives were harsh compared to the American 'engineer' and his crew, who, from the mid-nineteenth century, had the benefit of a glazed cabin house that provided good forward visibility, unlike the portholes of European locomotives. With advances in signalling and increased speeds, even the European trains eventually sheltered the driver's compartment, which evolved into a comfortable closed cab only with the production of diesel locomotives in the post-war period.

Hollywood repeatedly portrayed the democratic virtues of the Pullman environment during the Depression of the 1930s, when the country was on the move in search of work, while at the same time realizing the American dream of total mobility. The 1932 musical *42nd Street*, choreographed by Busby Berkeley, featured an elaborate kinetic stage set, designed by Jack Okey, and based on a typical Pullman Palace Car, which divides longitudinally to reveal the aisle and sleeper berths occupied by an assortment of Runyonesque characters in night clothes singing the song 'Shuffle Off to Buffalo'. The episode presents the detailed portrait of a culture that arose as a specific response to the convertible design of the Pullman Palace Car.

To Niag'ra in a sleeper
There's no honeymoon that's cheaper
And the train goes slow.

Off we're gonna shuffle
Shuffle off to Buffalo

For a little silver quarter
You can have the Pullman porter
Turn the lights down low

Off we're gonna shuffle
Shuffle off to Buffalo

It was in the 1940s that long-distance travellers, particularly in America, last accepted the communality of a Pullman sleeping berth, before the compelling grip of private automobile space entirely took over their concept of overland travel. Before jet airliners first completed the transcontinental journey in five hours, and the car provided independence and solitude on shorter journeys, the curtained sleeping berth offered simple comforts at the expense of insulation from other people.

## Bullets and Beyond

From the 1960s it became apparent to the world's railroads that competition with the airlines for passenger traffic would demand a major rethink of their services and design. For long-haul travel, the airlines were supreme. High-speed inter-city trains, however, could compete effectively with the airlines in terms of door-to-door travel times, for journeys up to around 400 miles, because of their ability to move passengers directly from city centre to city centre, without the inconvenience of the long check-in times required at airports.

Trains capable of cruising at 100 mph or more needed dedicated infrastructure, however. This was first demonstrated convincingly in Japan, where the 320 miles between Tokyo and Osaka could be covered in three hours on new track used only by the revolutionary high-speed Shinkansen ('Bullet Train'), inaugurated in 1964. For the first time, drivers had at their command an electronic telecommunication and information system, displayed on the cab control console, which replaced traditional line-side signalling. For the passengers, efficient airline-style accommodation supplanted the spacious civilities of earlier de luxe trains.

The interiors of the fast and efficient Spanish AVE intercity trains are relatively conventional in terms of style, but their excellent detailing and amenities, such as a spacious café car and duty-free shop, offer passengers a little diversion and an opportunity to get up and walk around. These are significant advantages when the main competitor is an aeroplane.

The three main attractions of these trains were their high speeds, city-centre departure and arrival, and the frequency of the service. Two classes of passenger travelled in rows of paired or triple seats facing the front, just as in an aeroplane, but with more leg and elbow room, particularly in first class. There were no private compartments, and the only special amenity on board was the buffet/dining car. The crew also served bento boxes to passengers at their seats, airline-style. The hi-tech Bullets were meant for only short-term accommodation, and their commuter-type interiors were relatively dull.

Although they pioneered modern high-speed rail travel, the original Shinkansen trains were unsophisticated when compared with more recent developments, both in Japan and elsewhere. Deutsche Bahn's ICE series trains have been at the cutting edge of developments since their launch at the turn of the millennium. As in the Shinkansens, their motors were located underneath carriages the length of the train, liberating space in the front control car for passengers. The ICE driver, seated in the nose of the train behind a deck containing instrumentation and controls, is separated from the passengers only by a clear glass screen through which they can see the track ahead, a sensational experience when the train is moving at nearly 200 mph. Although apparently innovative, the front observation lounge was pioneered in a control car of 1952 for the Italian Breda Company. Designed by the architect Gio Ponti, Breda's Belvedere car featured a spacious and luxurious observation lounge occupying the fully glazed nose of the train, underneath the driver's cab and providing a fantastic view for the speed-loving traveller.

While offering a thrill to those seated near the front, the interiors of the ICE trains do not waste space on specialized facilities and are designed to serve the requirements of the overwhelming majority of passengers, who travel for business. These trains, and very similar models used on the French TGV and Spanish AVE high-speed lines, are like huge mobile executive offices. Seats are ergonomically contoured armchairs with fold-down writing desks of the aeroplane type, but large enough to be useful for a traveller with a laptop computer, papers and coffee. These executive loungers also offer occupants power sockets, fax and Internet connections. Acoustic privacy is available in compartments set aside for meetings. first-class passengers are accommodated in more spacious interiors, some with wood-trimmed, leather-upholstered chairs, as in a Mercedes or Lexus. These are arranged in a 1 + 2 configuration across a

The French train operator SNCF commissioned MBD Design, working in collaboration with the fashion guru Christian Lacroix, to redesign the interiors of their TGV trains, using sleek forms and stylish colours reminiscent of Braniff airliners of the 1960s.

central aisle. All passengers have access to video/headphone entertainment consoles familiar from modern airliners, as are the train's high-quality finishes and branded colour schemes.

Other high-speed trains have been designed to appear and to feel as place-like as possible, while still offering economically viable seating density and a limited range of specialized amenities on board. The interiors of the Italian ETR500 were designed to create a luxurious atmosphere for travellers through the use of strongly coloured materials, style changes in separate areas within the cars, contrasting textures, and sensuously curved walls and ceilings.[36]

The French railway operator SNCF employed Christian Lacroix, with seat designers Compain / MBD, as consultant for the €270 million refurbishment of its fleet of TGV trains in 2003, with the aim of introducing a note of Gallic chic to the interiors of the carriages. The sleek prototype featured asymmetric seats upholstered in bright plum, orange and lime-green fabrics reminiscent of 1960s Braniff airliners. Could this be the end of the plain train? Probably not! The version that reached production was upholstered mainly in grey.

Even such purely cosmetic efforts, to generate a sense of being 'somewhere', are rare. With the near total withdrawal of specialist facilities or amenities on board modern trains, they fall increasingly into the category of efficient, high-density, mass transporters. Since safety considerations associated with increased speeds will ensure that secure seating remains a priority, any opportunities to walk around the train freely will be reduced to the minimum as they are in current airliners. 'Please keep your seatbelts loosely fastened throughout the flight.'

A survey of potential rail passengers in Northern Ireland in 2003 found that a cross-section of the public was more concerned about health and safety issues than entertainment or even the comfort of the seat in which they travel. Issues included smoke-free carriages, CCTV on-board security systems and access for disabled passengers. Designers for companies such as Translink of Northern Ireland work in teams and in collaboration with a variety of international consultants to detail designs in response to the expressed interests of specific passenger groups, and principally the powerful disabilities lobby.[37] Considerations of style have meanwhile diminished to choices of colour and a narrow range of fireproof materials, also determined by regulators and by surveyed expectations of average customer preferences. As a result, wherever you go, there are few visual surprises inside the modern railway car.

It may be surprising, nevertheless, how directly the traditions of nineteenth-century patent furniture apply to the design of today's seating for disabled passengers. Wheelchair access was an aspect of transport design

Natural, lightweight materials, evidence of traditional craftsmanship and striking graphic art works complement the sleek internal spaces of the latest generation of Japanese high-speed trains. Such features give these cars a look in keeping with popular concerns for the environment. These interiors state clearly that train travel is the attractive way forward for mass transportation in an era of global warming.

and of design in general that received almost no attention before the 1970s, when the global acknowledgement of minority rights led to an avalanche of regulations and guidelines to support those newly established rights. The British design consultancy Priestman Goode, working for Virgin Trains, designed wheelchair-accessible seating with a flip-down, theatre-style seat and a height-adjustable tabletop, both of which carry all the hallmarks of the convertible furniture designed for passenger comfort by Pullman and his contemporaries.

According to current ecological thinking, passenger rail travel in the twenty-first century should concentrate on regional inter-city routes. There will be less call for overnight services, because routine trips of more than three hours duration will commonly be handled by aircraft. Trains of the future will probably be more or less like the transporters of today, but faster, safer and more comfortable in an ergonomic sense. Passengers could become accustomed to viewing the passing landscape at speeds up to 400 mph.[38] Such speeds are being achieved by new magnetic levitation trains (MAGLEV) on the experimental route in Emsland, Germany, and in the world's first commercial MAGLEV service in Shanghai. However, slow-moving overnight services will continue to be provided on nostalgic cruise trains where passengers can experience the luxury and romance of train travel, as it was in the 1920–60 period, on restored, antique rolling stock, equipped and staffed to the highest expectations of the well-heeled tourist.

## Metros, Tubes and Subways

The experience of public transport in any of the large cities of the world reveals certain significant characteristics of the culture in that particular place, even if the system used, bus, underground railway, tram or taxi, has much in common with the corresponding system in other parts of the world. Riding one of the colourfully painted trams of Melbourne quickly conveys the vivacity of that seaside Australian city. The splendour of the Moscow underground speaks clearly of the aspirations of Soviet metropolitan culture. And a historical excursion around London Transport design tells a revealing story about British modernism.

The history of London Transport is a case study often cited in the literature of design because it exposes so many issues regarding the aesthetic life of the community and the history of corporate identity programmes. Some writers, such as Nikolaus Pevsner and Adrian Forty, have considered the interior design of buses and underground trains, as they were influenced by the singular design ethic pursued in the 1930s by London Transport. Yet more can be said about LT interior design as compared with recent developments in public transport.[39] London is uniquely revealing as a pioneer of the major forms of modern transportation, and the city benefited from excellent transport design and coordination during the inter-war period. Later, however, it plunged into transport chaos caused by lack of investment and imagination. It is now a useful mirror on changing attitudes towards the functions of public transport in the twenty-first-century metropolis.

London had pioneered the omnibus and ran the first metropolitan underground train service in the world. By 1907 nearly all of London's underground trains were electrified, allowing the construction of longer and deeper tunnels than those permitted by the steam locomotives of the previous century. At the same time, the many private companies, which owned the independent underground railway lines that operated across the capital, realized that their survival was dependent on cooperation to make it possible for passengers to transfer easily from one line to another, enabling them to reach any destination with ease and economy. By 1913 the United Electric Railways of London (UERL) had acquired most of the independent underground lines and the largest bus company in the city, the London General Omnibus Company (LGOC), creating a single company controlling most of public transport in London.

Albert Stanley (later 1st Baron Ashfield), the far-sighted director of the UERL, realized that the company's image was severely compromised by the wide variation of rolling stock, station design and other visual aspects of the system. The service also suffered from the conservatism of Scotland Yard, who in the 1920s set the specifications for many aspects of the company's transport vehicles and slowed their development. Therefore, Stanley advocated the institution of a powerful autonomous monopoly, for the smooth management of the company and the welfare of its employees, and especially for the benefit of those who used the service.

The culmination of this process of amalgamation was the formation, in 1933, of London Transport, which set out to create a recognizably integrated public transport system for all of London. Frank Pick was the extraordinary manager, who recognized that art and design had importance to the company well beyond the sphere of pure aesthetics. As Adrian Forty put it, Pick recognized that design could 'affect the entire way in which the population of London regarded London Transport and its influence on the development of ideas about the size, shape and character of the city'.[40]

Christian Barman, London Transport's Head of Publicity in the 1930s, called the company a 'civilizing agent', and Nikolaus Pevsner described how this effect was managed by Pick: 'to Pick art was always a means to an end', an ongoing process of taking 'patient steps forward' to improve all aspects of the service. Regarding the interior design of Tube and bus stock as it evolved under Pick's leadership, he pointed to engineering advances, such as the shift from timber to metal construction, and improvements in seating layout and details such as the piers between the windows of Tube trains.

> By means of great technical ingenuity they too have been made flush. Visible screws have been all but abolished. The alternation of long-wise and crosswise seating is introduced in order to obtain the widest standing accommodation by the doors and taper it down away from the doors.[41]

Thus, engineering and the practicalities of passenger accommodation motivated design innovation, from the major considerations of seating layout to the smallest details of hardware, which also advanced the aesthetic aspects of the service and enhanced the reputation and performance of the organization.

Architecture, uniforms, poster design, typography, station furniture, ticketing machines and all other aspects of the transport environment, designed in the years from 1933–9, were conceived with a singular intention: to present London Transport as a progressive, rational public service of the highest quality. Corporate identity was meant to encourage more travel by presenting an image of ease, efficiency, speed, safety, cleanliness, comfort and pleasure. While the constituent elements of the system, from infrastructure to ephemera, were important to achieving this impression, it was the vehicles that were at the heart of the passenger's experience. Although the external appearance of Tube trains, and even more so buses, was important, it was the interiors of the vehicles with which passengers interacted most intimately.

Tube trains had been evolving steadily for seventy years before the formation of London Transport, and the incremental improvements had succeeded by the mid-1920s in creating a comfortable environment for passengers. Yet the generations of carriages introduced in 1923 and 1927 still required large compartments behind the driver to house their bulky motors. Carriages featured a 60-year-old Pullman-style cross-section, with a raised roofline over the aisle and a deep coving above the seated passengers.

Still influenced by Victorian and Edwardian taste, the cars were finished with heavy, darkly polished woodwork on their end walls, interconnecting carriage doors, window frames and as accents around advertisements and information posters. Gooseneck lamps with frilly glass shades provided dim lighting. Although the longitudinal bench seating of the 1923 stock was upholstered in leather, it offered no separation between individual passengers' spaces, and it was drab. The cars of 1927 were slightly better: they featured cantilevered arms between individual places on the long seats and introduced the colourful, moquette upholstery that became a trademark of LT vehicles thereafter.

With the dramatic expansion of the underground lines to the north, west and south of the city centre and into the rapidly expanding suburbs, LT had the opportunity to create a thoroughly coordinated look to all new elements of the service. The architect Charles Holden developed a formula for the design of new stations, using a standardized palette of modern materials and a set of distinctive forms, which gave all new LT buildings (and those being renovated) an easily recognizable identity, although each was designed individually in relation to its site. Stations such as Cockfosters, Southgate and Arnos Grove, on the Piccadilly Line, were characterized by sweeping curves or sharply angular geometries, high and airy booking halls with large areas of glass – to admit the maximum light in daytime and to illuminate the façades dramatically at night – and the expressive use of concrete, brick and steel to create elegant entrances to the underground system.

The exquisite detailing of these stations was evident, for example, in the attention paid to the escalators, which conveyed passengers from the booking halls to the platforms below ground. At Southgate Station, the gracefully arched escalator shaft was detailed with discreet tiling in LT colours and softly illuminated by tall bronze uplighters, set in rows between the sweeping forms of the mechanical staircases and continuing into the hall beyond. The effect was grand and calming, luxurious and diverting. On the station platform, travellers were treated to a gallery-like display of large advertising posters, many of which were sponsored by LT to promote its services. Modern artists, such as Edward McKnight Kauffer, were enlisted to portray attractive destinations in and around London. Their posters stimulated desire to travel for pure pleasure and entertainment in the years before universal car ownership, while they also distracted attention from the tedious grind of commuting to work.

Although many older stations remained unchanged during the 1930s, the arrival of one of the new trains at any station platform instantly communicated the image of a modern transport network. The Tube stock of 1938, designed under the direction of W. S. Graff-Baker, represents the high-water mark of London Transport design. These sleek red cars abandoned the clerestory-roof profile of earlier models in favour of a smoothly arched roof, giving the trains a practical, streamlined shape, which was also designed to facilitate the operation of newly automated train-washing machines. Relocating all motors to the underside of the carriages freed the whole interior of the train for passenger space. London Underground's tradition of prioritizing seating was continued, with separate sections offering the choice of face-to-face booths, comfortable for small groups of people, or longitudinal seating, positioned above the wheel arches that protruded into the space of the car.

Like the new buses of the period, the interior surfaces of the trains were designed to appear seamless, their cladding panels smoothly contoured, all junctions and fastenings concealed to give them a look of modernity and high quality. Metal panels were painted in a soft green, harmonizing with bold, geometrically patterned upholstery, which was robust and luxurious to the touch. LT commissioned textiles, to their own very specific briefs, from prominent modernist artists and designers, including Enid Marx, Marion Dorn and Paul Nash. Marx wrote about the briefing she was given before designing her first upholstery pattern for LT:

> Barman . . . started off by informing me that some of the trains started in the country, in daylight, and then went on underground, and this fact meant that careful consideration had to be given to

A London Underground train interior of 1938. These trains were mass transporters of quality and comfort, their tubular spaces differentiated by half-walls, standing zones near the doors and varied seating arrangements. The quality of details and finishes was similar to those of the period's high-class motorcars.

the colour or colours used in the interiors of the trains . . . as certain colours responded well to daylight but not nearly so well when going underground into artificial light. Because of this, he had chosen green as his main colour for the seating materials . . . you had to remember not to produce a design that could in any way be dazzling or give the passenger a feeling of sea-sickness . . . the variation of class of customer was very wide. You got dustmen and people doing outside and dirty work going by train from job to job, you also had ladies in fine clothes going to parties and not wanting to sit on a dust heap.[42]

Understanding their clientele enabled LT bosses to guide talented designers to produce the most appropriate work for the purpose, from architecture to the smallest features of the vehicles' interiors.

On board the 1938 trains, bright, even lighting was provided by stylish Art Deco 'shovel shade' lamp fixtures, fitted in series to the coving above the heads of the seated passengers and creating reflective highlights in the polished surfaces of the wood and metal work. Woodwork ceased to dominate the interiors, but remained as trimming around the windows and covering the guard's control panels. finished in a light stain, the woodwork combined with chrome framing around the vent windows to create an impression of high-quality craftsmanship and attention to detail, as would be found in a Bentley. The warm colours, luxurious fabrics (trimmed in leatherette), hardwood floors, neatly integrated advertising panels, clearly presented Tube maps and other LT information panels all induced a sense of familiarity, comfort and diversion for passengers.

The palette of materials also had acoustical advantages, softening the mechanical noises of the train and dampening the sounds of other passengers' conversations and movements. Details, such as the moulded Bakelite handgrips for standing passengers, the armrests between seats and the mechanical window latches, were all smoothly contoured and pleasant to handle. The backs of the seats were subtly curved to provide physical support, and also to give each passenger a visibly separate space, even in the opposite-facing double benches, which had no central armrest.

Subsequent generations of London Tube trains have refined and updated the 1938 design. Aluminium construction replaced the wood-framed bodies of earlier trains and external paint was eliminated, saving cost and weight. Inside, by the late 1950s stainless steel and rubber replaced painted metal for grab poles and wall cladding, while in the 1980s grooved, rubber flooring supplanted hardwood. Glass area increased somewhat, but seating patterns remained essentially the same, although on some lines standing passengers were provided with upholstered rests against which to lean. Longitudinal strip lighting further brightened the interiors and, eventually, acid-coloured grab poles replaced stainless steel to improve their visibility for partially sighted people. Yet the padded seats covered in moquette upholstery remained.

## The Great British Bus

In spite of the growing underground rail service at the turn of the twentieth century, London's streets were clogged with horse-drawn carriages, omnibuses and cabs, in addition to all the goods vehicles required to service the city's shops, offices and manufacturing works. For most of the residents of London, the horse bus was king of the road, but its regal status was under threat from the growing realization that the internal combustion engine was more efficient and economical than the horse. The last horse-drawn bus ran through the streets of London on the eve of the Great War in 1914, but its days had been numbered since 1908, when the LGOC put into service its first reliable motor bus, the Type B, the first bus in the world to be mass-produced. The Type B quickly dominated London's established routes and demonstrated to passengers its relative speed and comfort, as compared with horse-drawn vehicles.

In the early days of motor transport, many different types of bus were on the road. Among them were some very elaborate machines with highly decorated bodywork containing grand interiors, finished with bevelled glass windows, etched mirrors, deeply tufted leather upholstery, metalwork detailed in the fashionable Art Nouveau style and highly crafted woodwork. Such luxury, however, was short-lived, as bus operators learned that the real issues for their business were reliability, economy and size. The capacity of the typical knifeboard omnibus was 26, while even the earliest motorbuses carried 34 passengers. Horses were expensive to keep and could not haul as heavy a load as a petrol motor, which was cheap to run, even with high taxes on fuel. The first motorbus routes in London were established in 1904 with around twenty vehicles, but by 1908 more than 1,000 were on the road. The LGOC, which was then the largest bus company in the world, had been experimenting with vehicles built by various manufacturers up to 1910, when they began building the Type B, the bus that put the company on its feet and that became the prototype for all subsequent London double-deckers.

Type B buses rode on hard rubber tyres, giving the passengers a similar jolting over the road as they would have experienced in a horse-drawn omnibus, but the smoother power of the internal combustion engine made the ride somewhat calmer for passengers, if not for the driver, who had to control a huge, primitive contraption in the full glare of public view. The enclosed lower saloon had large fixed windows with slim hinged vent windows above them. Seating was along the sides of the body, facing inward, just as in the earliest omnibuses of the previous century. Seats were upholstered in a subtly patterned woven

The evolution of the London bus: the open-topped Type B of 1908, the NS of 1925 and, the most successful of all, the RM (1959–2005). The most influential head of London transport, Frank Pick, impressed a unified design ethic, underpinned by a strong sense of social responsibility, on the entire transport network of the city.

fabric, a curving brass rail dividing each of the two long benches in half. There was also space for standing passengers in the aisle, with grab-rails suspended from the ceiling, front to back. Inside, the craftsman-like wooden structure of the body was painted white above the backs of the seats and stained below. With electric lamps fitted to the ceiling in purely functional clear glass holders, the bright interior appeared comfortable and welcoming, like a small, well-appointed waiting room.

The stairs were another matter. They were unprotected from the weather and so were wet much of the time, making the ascent to the upper deck perilous, and the descent even worse. Yet in dry weather the top of the bus was the more amusing place to sit, to show-off and to flirt, out in the open air, high above the pedestrian traffic on the pavement. Upstairs, passengers sat in pairs, facing the front on slatted wooden seats, like garden furniture, flanking a central aisle. The railing of the external stair continued around the perimeter of the top deck, preventing passengers, and especially the more high-spirited, from tumbling overboard while moving to or from their seats. The combined effect of the Type B's decorated paintwork, brightly coloured livery and advertising posters pasted to the exterior was of a giant fairground object, parading through the streets of the city.

During the war of 1914–18, many London buses were converted for military service, the Type B earning the same patriotic status as the heroic Paris taxis, which ferried troops and supplies to the battlefront in the defence of the city. Many thousands of British and Allied soldiers rode in Type Bs, hastily converted as troop carriers, their windows removed and boarded over or covered with rush matting. Some were lightly armoured and others were camouflage-painted, as seen in photographs of troops being driven to the battle of Ypres. Other pictures, however, show buses with their original bright paintwork and advertising posters still in place – incongruous, carnival-like, joy-ride vehicles in the midst of grim death, carrying British troops into battle on the Belgian Front. Whatever the external appearance of these machines, the interiors must have been frighteningly dark and claustrophobic for the young men who rode them into the gruesome theatres of trench warfare. Solid rubber tyres and springs designed for city streets could only add to the discomfort of the troops over the rough and rutted roads of the Western Front.

The first new LGOC bus to appear after the Armistice was the 54-seat S Type, which broke significantly with the traditions of horse-bus design. From the perspective of comfort, however, the S was little better than the B, mainly due to Scotland Yard regulations against such advances as covered top decks, windscreens for drivers and pneumatic tyres, all of which were considered dangerous. Yet the interior of the lower saloon on the S

Type was markedly different due to its lower platform and wider body, which allowed all seats to face the front, except those above the rear wheel arches. Now, most passengers could sit facing the direction of travel and enjoy a good window view.

By 1926 the police were ready to permit the enclosure of the top deck and other innovations that significantly improved the comfort of the passengers and crew, yet the buses built throughout the later 1920s and early 1930s were still close relations of the old Type B. Inside there remained visible evidence of their wooden structural framework, and the junctions of all the small individual elements of timber construction were plain to see.

The real breakthrough to a new generation of buses, consistent with the corporate identity of London Transport, was the design of the RT bus, commissioned by Frank Pick and introduced in 1939. This was a sleek, modern

vehicle. Its handsome proportions were the aesthetically considered result of its fitness for purpose, with flush body panels, radius curves joining all the major planes of the body, flush window frames and highly refined details throughout. The inauguration of the RT marked the beginning of the second phase of London bus history, one that would last into the new millennium. Along with the Tube stock of 1938, the RT represented the pinnacle of London Transport's functionalist approach to design at the end of the period of high modernism, just before the Second World War. The resulting product was so successful that its basic form was retained for the revised RM (Routemaster) model introduced in 1959, the last of which retired from normal service in 2005.

The upper and lower saloons of both the RT and RM variants provided 64 seats with an additional eight standing places. They were finished in

The timber frame, clearly visible in the saloon of the Type B bus (above), reveals its genetic link with the horse-drawn vehicles of the previous century. In an RT bus (below) the seamlessness of curved metal panels and uninterrupted ribbons of glass represented the cleanliness and order of the modern mechanical world.

smooth metal panels, their junctions and connections to the frame completely concealed. All corners were curved, creating a seamless, smoothly enveloping space around the two rows of tubular steel seats, upholstered in the LT moquette fabrics. The polished metal frames of the seats provided a clear demarcation of personal space and joined elegantly with the vertical grab poles. Each light bulb was recessed into the flat surface of the domed ceiling, creating two rows of illuminated hemispheres above the passenger seats. Even the organically formed window winders contributed to the sense of harmony and quality apparent throughout the vehicle.

Although RT and RM buses resemble each other closely, they were very different in terms of construction and mechanics. All buses up to the RM had been timber-framed. Even chair frames were made mainly of timber until the RT introduced patented tubular steel seats. The RM, the first London bus using an entirely light alloy structure, was designed under the direction of William Durrant, London Transport's Chief Mechanical Engineer, and styled by the industrial designer Douglas Scott. The Routemaster was very advanced at the time it was put into service in 1959. The driver's compartment was ergonomically designed and introduced, for the first time, automatic transmission, power steering, hydraulic brakes and independent suspension, making the RM easier to drive than most British cars of the day. Yet to the passengers, both the RT and RM buses exuded an impression of modernity and quality, and, above all, their interiors provided the physical comfort of a good motorcar.

With an economic boom, an exponential rise in car ownership and changing values regarding the nature and purpose of public transport since the 1960s, the design approach of Pick and Holden has been replaced by pure economies of scale. Although at present a ride on the London Underground remains more or less as it was in the days of Pick and Holden, with train interiors retaining the basic qualities established by London transport in the 1930s, on modern buses all is foreign. New buses in London are imported from Sweden or Germany, and apart from their red paintwork they bear little resemblance to the elegant and comfortable vehicles of the past.

Eighteen-metre-long Mercedes Citaro single-deck articulated buses have Green credentials, are wheelchair-friendly, and are successful in moving up to 159 passengers, 120 of them stand-

Twenty-first-century buses are spartan affairs, dedicated to moving large numbers of standing passengers at the lowest possible unit cost per mile. The Mercedes Citaro artec, seen here in service with the Barcelona transport authority, is finished in shades of grey throughout its mechanistic interior space. This is a no-nonsense mass transporter!

ing, as cheaply as possible over relatively short distances. Consequently, the interior appears as a dense forest of grab bars, which creates a cage-like environment. Seats of hard moulded plastic perch at staggered levels, while the accordion vestibule, at the junction between the two halves of the body, is the unstable centre of the vehicle, as it bends around corners and curves.

Echoing the tradition of the London bus, externally these machines are decorated to project a jolly image through their bright red livery and colourful advertising posters. Unlike the sophisticated interiors of earlier London buses, however, the Citaro interiors are finished in a riot of contrasting greens, blues, yellows and oranges covering their wildly patterned upholstery, floor coverings and interior hardware, resulting in a cacophony of quickly soiled surfaces – like the interior of a detention cell for delinquent pre-schoolers. This is the choice of the client, rather than the manufacturer. Identical M-B Citaros in Barcelona are painted a subtle cream with red accents, the livery of the Barcelona Transport Authority, while their interiors are finished and upholstered in solid greys with brushed stainless-steel grab poles. Floors are dark grey linoleum. The result is a no-nonsense environment that expresses fearlessly the true, mechanistic character of this particular type of mass-urban transporter. The contrast between the interior styling of Citaros in London and Barcelona demonstrates the futility of loading a stark transporter with the superficial imagery of pleasure and diversion. It also illustrates the flaw in globalised transport design – the Citaro's utilitarian design is appropriate on the short routes of a small city, such as Barcelona, but fails in London, where the passenger may be on board for more than an hour.

## The Coach and the Caravan

The first motorized public vehicle designed specifically for joy riding was known as a charabanc. These were large open touring cars, seating 20 to 30 people. Many of the earliest were converted war vehicles, which became particularly popular in Britain, where they achieved their greatest success providing working-class holiday outings in the 1920s and '30s. Charabancs attracted a dubious reputation as rolling bar rooms, but were much loved by those who enjoyed packing into their heavily padded, tufted-leather bench seats for open-air rides to the seaside, lake or river where tourist facilities were beginning to spring up between the wars. The charabanc became an icon of liberation for working-class people, who for the first time had the luxury of a paid holiday in a motorcar.

In North America, the long-distance bus attained iconic status through its roles in a canon of novels and films about the independent spirit and the lure of the highway. Although many companies had run inter-city and coast-to-coast bus routes since the 1920s, the Greyhound Bus Lines came to stand for low-cost, long-distance travel in the US. Among their best-known vehicles was the Scenicruiser, developed by Raymond Loewy and the General Motors styling department over a ten-year period and launched in 1954. Its sleek, horizontally ribbed aluminium body, with lines sweeping eagerly forward, suggested fast, state-of-the-art road transport for those travelling on a budget or for anyone living far from a train station.

The 50-passenger Scenicruiser featured a split-level design, with the driver and some passengers seated on the lower level at the front. One third of the way back, the passenger compartment was raised above a large luggage hold, with a second windshield lifting the roofline to give the upper-level passengers an unobstructed view forwards. Loewy's extensive design research resulted in significant changes to the space, amenities, internal layout and seat design of the long-distance bus. Greyhounds offered the maximum comfort available on the road. Amenities included well-planned lavatories, vending machines, reclining seats and large, 'panoramic' windows to provide the best possible views of the passing scenery. Yet it could never offer the space and comforts of the train nor the freedom and privacy of the car, its two main competitors, and so it remained a second-class form of transporter.

Nevertheless, the Greyhound became the stuff of legends. Beyond its function of moving passengers cheaply, the interior of the long-distance bus proved to be an ideal space for musical performance, exploited in its role as a tour vehicle transporting bands and orchestras to concerts throughout the world. Many films, including the pre-video 'soundies', forerunners of today's music videos, were set inside buses, where band 'rehearsals' and spontaneous jam sessions became the subject of the film. Memorable performances by various popular orchestras and singers, such as the young Ella fitzgerald, were recorded in this way before the Second World War, while later films, such as *New York, New York* and *Lady Sing the Blues*, the biopic of Billie Holiday, exploited the linear space of the aisle for performances, while the paired seats became the arena for romantic and erotic encounters stimulated by the music.

In the opening sequence of the film *Midnight Cowboy* (1969), the small town southern boy, headed to New York to embark on a career as a gigolo, quietly contemplates his past and excitedly anticipates his future while watching the passing landscape through the streamlined window of a Greyhound. Trying to use his rustic charm on fellow passengers, who come and go, indifferent to him, he learns the rudiments of becoming a classic American loner, displaced and incongruous in the rootless life of the highway. The film both begins and ends with a bus ride, the interior of the vehicle, its close-coupled, high-backed seats and large window views portrayed as both alienating and intimate, a conveyor of dislocated hopes, of psychological as well as physical discomfort brought on by prolonged immobility. Its arrangement of narrow, closely spaced seats forced strangers together, intimately, and simultaneously ensured the isolation of the lone traveller.[43]

The camper version of Volkswagen's Microbus represented a popular desire to return to nature. It was the first compact and economical mobile home, and it became one of the most recognizable symbols of the freedom-seeking alternative society of the 1960s.

Whereas the Scenicruiser was a sleek example of mainstream industrial design, in the aesthetic free-for-all of the 1960s long-distance bus travel was given a different twist, reported by the journalist Tom Wolfe in his book *The Electric Kool-Aid Acid Test* (1968). Wolfe told the story of the novelist Ken Kesey and his retinue known as the Merry Pranksters, who signified their hip, druggy way of living through a fashion and graphic style inspired by the visual effects of the hallucinogen LSD. The main icon of the Pranksters' image was an old International Harvester school bus adapted, ad hoc, for group travel with bunks and a powerful music system. The most inventive aspect of the design was the lurid, psychedelic mural with which the Pranksters covered both the exterior and interior of the bus, expressing the shifting proportions and intense colours of the acid experience. In it they drove relentlessly over the United States spreading the message of sex, drugs, and rock and roll. The destination board said 'FURTHER', suggesting psychological as well as geographical progress.[44]

Kesey was well known for his novel *One Flew over the Cuckoo's Nest* (1962). He also formed a highly publicized connection with the Beat Generation writer Jack Kerouac, whose literary model, Neal Cassady, drove *Further* from California to the New York World's Fair in 1964. Wolfe reported on the trip, and the bus was photographed for *Life* magazine. Consequently, its influence was quickly felt in the 'psychedelic' interior decoration of student apartments, hippie pads and in the home-made interior decor of countless vw Microbuses used for nomadic voyages of discovery in the heady days of the alternative society, during the 1960s and '70s.

Equally innovative for its time, if more conservative in design, was the prototypical motor home built in 1952 by a California aircraft manufacturer as a weekend getaway vehicle for himself and his wife. Otto Timm designed his self-propelled house on wheels, and had it constructed using lightweight plywood technology, pioneered for Second World War air-

The modern motor home began with unique experiments by self-builders using the latest technology to suit new social and domestic requirements. The house-on-wheels was a centuries-old dream realized by a massive mobile population in the second half of the twentieth century.

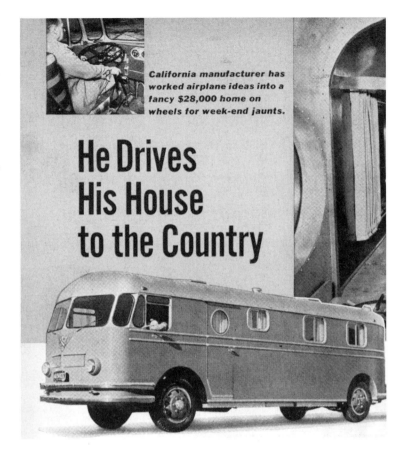

California manufacturer has worked airplane ideas into a fancy $28,000 home on wheels for week-end jaunts.

# He Drives His House to the Country

craft production. The interior was elegantly panelled in mahogany veneer and featured convertible, collapsible, foldable and sliding furniture, in the tradition of Pullman, to make the most of the interior's 28-foot length. The high specification included a kitchen equipped with refrigerator and compact cooking appliances. There was also ample closet space and full air conditioning. Timm was a clever independent thinker, who brought together in his innovative design the domestic comfort of a house trailer (caravan) with the relative compactness and easier handling of a bus.[45]

From the 1960s to the end of the century, companies such as Winnebago developed the motor home commercially, and it eventually superseded the car and house trailer combination as the standard type of holiday vehicle in America and around the world. It encapsulated the historical dream, sought by the royal owners of Berlin carriages in the eighteenth century, of travelling fast and enjoying new spectacles at every bend in the road, with all the comforts of home and total independence. It also had the advantage of comprising a single space for driving and occupation, as opposed to the car + trailer combination.

Chic interiors by Christopher Deam have brought the classic American Airstream trailer, produced since the 1930s, into line with the technical and aesthetic expectations of the twenty-first-century camper. Ultra-lightweight materials and fixtures enabled the designers to include all the amenities of a good modern hotel room in the space of even the smallest Airstream trailers.

The house trailer, however, remained an important feature of family leisure in the car culture of the twentieth century. The Airstream trailer, conceived by Wallace Byam and produced in various sizes from 1936, became one of the most easily recognizable and attractive icons of American modern design. Unsurprisingly, the inventor / architect Buckminster Fuller owned one. Inspired by aircraft construction, the riveted aluminium skin of the Airstream trailer was polished to emphasize its streamlined form. Inside was an economically planned and cleverly fitted plywood-panelled room with leatherette upholstery, much like the cabin of a modern yacht.

In the 1980s and '90s the Airstream company somewhat lost its way, fitting their trailers and motor homes with interiors in heavy, traditional styles. However, with the recently revived fashion for mid-twentieth-century contemporary design, the company commissioned Christopher Deam to create interiors for its new models, which reflect the functionalism of the pre-1970 Airstreams. The new interiors feature exposed aluminium wall surfaces, clean-lined furniture, boldly coloured laminates and smoothly integrated equipment, including stainless-steel cooking appliances, flat-screen television and Internet connections, proving the durability of the original Airstream concept and design.

Despite some genuine design credentials, the house-trailer has always attracted a dubious reputation and was parodied most spectacularly in a feature film of 1954 starring the popular television comedy couple Lucille Ball and Desi Arnaz. *The Long, Long Trailer* was the story of a cross-country holiday plagued with misadventures, in which the couple live the full American Dream of unlimited mobility and ultimate comfort in a lime-green and turquoise 40-foot long trailer attached to the back of a new Mercury convertible. Karal Ann Marling described the trailer's

> colourful interior, replete with built-in sectionals and appliances . . . the stove with the glass picture window in the door . . . matching sets of pastel towels and kitchen gadgets . . . a deep freeze and a TV set. The latter . . . is the telling detail that will make a rolling house into a true home.[46]

## Wrap-Around Visions

'Now all we have to do to enter the realm of art is take a car'[47]

Lucy's and Desi's convertible was, itself, their living room on wheels. It was the glamorous leather-upholstered, colour-coordinated setting where they chatted, spooned and planned the day-to-day details of their excursion. The open-top car harks back to the tradition of the nineteenth-century carriage, described by Baudelaire and painted by Guys and Degas, who portrayed the upper classes parading their wealth and finery in their open carriages at the Longchamp racecourse. The modern convertible, by the mid-twentieth century, had established a particular display function, linked to the tradition of open horse carriages, and particularly to the phaeton, which had provided the inspiration for a turn-of-the-century French music hall ditty: 'Gaston likes to lay it on / In his horseless phaeton'.[48]

The horse-drawn phaeton and the type known as a buggy provided blueprints for the early motorcar. Because of the low power output of early internal combustion engines, the first cars had to be light and, therefore, carried very little bodywork. They had no interiors and were ridden on, rather than in. At first, clothing stood in for bodywork. Drivers and passengers protected themselves from the elements by wearing enormous 'duster' coats, veils, long-visored caps, gloves and goggles. As more powerful engines were developed, speeds increased, and the technology of body design advanced, the addition of a windscreen, roof, doors and side windows became possible and desirable to all those who wanted to use the car as daily transportation.

Specialist clothing protected drivers from the elements in the days before enclosed car bodies. British advertisement, c. 1900.

The fully closed car was an established part of the motoring scene before the first World War, but at that time it was seen as a vehicle of privilege, while ordinary folks still travelled out in the open or under a folding canvas roof. Henry Ford put an end to this situation by introducing by the later 1910s several reasonably priced, closed models – coupé, sedan and wooden-bodied station wagon – providing the average motorist with a carefully appointed and upholstered interior.

Alfred Sloan, the President of General Motors, also recognized the importance of the closed car to increase the utility, comfort, desirability and sales of automobiles. He employed all the resources of design and construction technologies and the new tool of styling to ensure that the six-passenger, four-door sedan became the standard all-purpose car, at every price level, by the late 1920s. And so, the open car became an alternative body type, a symbol of personal display and glamour. It belonged on Sunset Boulevard and anywhere else that posing-power was an asset.

With the ascendancy of the closed car, the convertible became a showpiece and a showplace, a setting in which the driver and passengers

# "The Plaid's the Thing!"

## THE CHRYSLER HIGHLANDER
### – the year's smartest car

FASHION'S newest note in the season's smartest car
F...the Chrysler Highlander, a pedigreed "Scottie" that
takes the blue ribbon for swank.

The plaid's the thing...and here
you have it in a car of year-'round
delight . . . a convertible Chrysler
that's ready for any kind of weather
at the crook of your finger.

A car with a *High-Torque* engine
that scales mountains or skims
boulevards with equal smoothness
and silence. A car with Floating
Ride that cradles you gently at the
center of balance.

**Push-Button Comfort**

Here's a convertible top you'll really use. To raise it,
you press a button; to lower it, you press a button.
Just that easily, you get protection or make the most
of a perfect day.

And on cold, raw days, you turn on the All-Weather
Aircontrol, which isn't just a heater, but an air-conditioning system (optional equipment at extra cost). It draws
in fresh air from outdoors...warms it...and circulates

**BE MODERN—BUY CHRYSLER!**

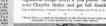

**WHY SHIFT GEARS?**
Touch the
throttle to go
. . . touch the
brake to stop.
That's Fluid
Drive. You
seldom use the
clutch or shift
gears. Transmits power to the
wheels through a cushion of oil
...no rigid metal connection. Positive, silent and liquid-smooth.

**FLUID DRIVE** *ONLY* $38 *EXTRA*

it in such a manner that there's a curtain of warm air
around all passengers. Air-pressure in the car is greater
than that outside, so no cold drafts can
creep in, front or rear.

You may have the Highlander with the
regulation coupe body, if you prefer. See
your Chrysler dealer and get full details.

*Fluid Drive is standard equipment on the Chrysler Crown Imperial.
Available at extra equipment on several other models.*

While the closed car became the standard automobile, the convertible found its place as a locus of haptic pleasure and extravagant display. The relationship between the fashion and car industries was particularly obvious when the interior was on public show.

could be seen enjoying themselves in the sun and open air. In this type of car, the design of the interior became all-important, since it was on show in the full glare of the sun and provided the colourful backdrop for the occupants and their clothing. For their cinematic rides in the Mercury, Lucy and Desi were elaborately costumed, she in close-fitting, brightly coloured dresses, her red hair aflame against the pastel uphol-stery, he in casual denim, which was then being marketed by Desilu Enterprises Ltd.

From inside the open convertible, the driver and passengers enjoyed an unlimited view above and around them – they could feel the elements as well as see them. Only the windscreen provided a frame or focus for the gaze of the occupants. In a closed car, however, the enveloping body-work defined very strictly what could be experienced or seen from the interior and how the view was divided into a variety of shapes. As early as 1905, the academic painter Hubert von Herkomer recognized the par-ticular sequence of images available to the aesthete driving in a closed car. He wrote: 'The pleasure [of motoring] . . . is seeing Nature as I could in no other way see it; my car having "tops", I get Nature framed – and one picture after another delights my artistic eye.'[49]

Other artists were soon exploiting the potential of the framed view provided by the windshield. Henri Matisse painted an early windshield view in his *Road to Villacoublay* in 1917. In this painting, the upright, rec-tangular windshield of an early closed car frames the central section of what is essentially a triptych, recalling Egg's *The Travelling Companions*, since there are glimpses of the roadside through the door glass to the left and right. We are shown parts of both the interior and exterior of the car, the steering wheel, horn, window-lifts, and the radiator cap, mudguards and brass headlights. Matisse painted from the back seat and propped another painting or sketch against the steering wheel for consideration or comparison. This interior scene brings to mind the many landscape views, seen through a framing window, painted by earlier artists includ-ing Fra Angelico and Vermeer. In *Road to Villacoublay*, Matisse represents the vanishing perspective of the road, the trees and the countryside as distinct from the architectural space inside his car, which protects him and privileges his relationship with the landscape. Matisse could take his studio anywhere.

Other artists pursued this viewpoint throughout the automobile age. Stuart Davis painted *Windshield Mirror* in 1932, employing his semi-abstract, hard-edged graphic style to represent images and events, simultaneously visible through the windscreen, side windows and in the rear-view mirror. A decade later, Edward Hopper recapitulated Matisse's subject and viewpoint but added the figure of his wife, seen sketching in the front seat of their mobile studio on a stop during a trip through Wyoming. Like Matisse, Hopper presents the space of the automobile at rest, like a genuine architectural space, but one whose location is not permanently fixed.

The Cubist influences in Davis's *Windshield Mirror* are also present in David Hockney's Polaroid photo-collage of the view through the windscreen of his Mercedes (*c.*1990). Hockney builds up a panorama by juxtaposing individual Polaroid photos taken from the driver's seat, but with each picture framing a small element of the overall view. The result is a startlingly detailed, wide-angle survey of the dashboard, of the view over the front of the car, and of the highway and desert beyond. Again, the interior–exterior relationship is the key to understanding the purpose of the work. The car is the laboratory for the exploration of this dualistic spatial experience.

However, the architecture of the car comes into its own most fully as a means of framing experiences when it is in motion. The first attempts to represent such phenomena were those of the Italian Futurists before the first World War. Giacomo Balla's many automobile paintings, completed between 1912 and 1914, included a series of pictures entitled *Abstract Speed* and *Cars + Light + Sound*, which represented, through composition, colour and line, the experience of movement – and particularly the new sensation of driving. The architect-filmmaker Chip Lord, who with Ant Farm built the *Cadillac Ranch* outside Amarillo, Texas, in 1974, communicated a calmer sense of driving in his video feature *Motorist* (1989). In this film, the central character, crossing the southern USA at the wheel of a vintage 1962 Thunderbird, tells the story of the car he is driving, while narrating and illustrating the colourful history of the Ford Motor Company. The juxtaposition of this cine-portrait and the background landscapes, framed by the windows of the car, emphasizes the car interior as both an observational space and a space for reflection, a rolling confessional, while also providing a visual inventory of the car's futuristic interior.

Other commentators have also represented the car interior as an architecture conceived for panoramic observation of the world in motion.[50] Gaston Rageot noted the way that kinetic architecture relates to a cinematic perception of the world:

> To see the landscape pass by a train or automobile window or to look at a film or computer screen the way you look out of a window, unless even the train or the cockpit become in their turn projection rooms . . . train, car, jet, telephone, television . . . our whole life passes by in the prosthesis of accelerated voyages, of which we are no longer even conscious.[51]

Such effects were greatly emphasized by the relationship of wall and window in the design of closed cars. The bodies of early sedans, such as the first generation of streamlined production models, beginning with the Tatra T87, designed by Hans Ledwinka, and Carl Breer's Chrysler Airflow, both of 1934, enclosed their passengers under a substantial, domed roof, supported by thick pillars, those at the front and back leaning in towards the vertical pillar at the junction between the front and back doors. General Motors dubbed their versions 'turret tops'. The result was a massively sculptural upper body shell punctured by relatively small, irregularly shaped window openings with rounded corners. These windows tended to be wide and low, contributing to the overall streamlining of the body and emphasizing the horizontality of the views streaking past the seated passenger. This horizontality became more pronounced over the years, even as the architecture of the roof became lighter and the glass areas increased.

As the design of American and American-influenced automobile bodies evolved in the 1950s and '60s, the functions of the convertible and the sedan or saloon merged somewhat. The amounts of glass area and the continuity of window openings increased dramatically after the Second World War. The advent in Europe of pillar-less coupés and sedans, advertised as 'hard-top convertibles' in North America, created open-air interiors under the cover of almost unfeasibly thin roof membranes, emulating the Plexiglas canopies of jet fighter planes.

In France, Citroën launched its futuristic DS model, designed by Flaminio Bertone, in 1955. The design of this car seductively embodied the aspirations of modern motorists through its emphasis on lightness and transparency. The hydropneumatic suspension system and wide, soft seats of the DS induced a sensation of floating over the road rather than riding on it. And the uninterrupted ring of glazing, liberating the roof plane from the cigar-shaped lower body, composed a spectacular, panoramic view of the world akin to the wrap-around vision of

Designed by Flaminio Bertone, the futuristic Citroën DS was driven from the most comfortable car seat of its time and offered its driver a dramatically advanced array of controls and information. The floating roof plane appeared to be joined with the lower body by little more than a surrounding curtain of glass, giving the passengers a panoramic view of the world. Bertone's design was both practical and poetic.

the Cinerama films that were being shown in the mid-1950s. It was also this car that attracted the attention of the semiologist Roland Barthes:

> It is possible that the Déesse marks a change in the mythology of cars . . . despite some concessions to neomania (such as the empty steering wheel), it is now more homely . . . the dashboard looks more like the working surface of a modern kitchen than the control room of a factory: the thin panes of matt fluted metal, the small levers topped by a white ball, the very simple dials, the very discreetness of the nickel-work, all this signifies a kind of control exercised over motion, which is henceforth conceived as comfort rather than performance. One is obviously turning from an alchemy of speed to a relish in driving.[52]

Both the text and the car are replete with Frenchness, demonstrating the persistence of national characteristics in both automotive design and its appreciation.

## Dead Man's Curve

Cars have become such an ubiquitous part of the lives of people all over the world that their particular uses are impossible to catalogue. Yet the 'homeliness' that Barthes points out in the design of the Citroën is a useful indicator of how we use our cars. The car interior has always been an alternative to home. It was and is the setting for the driver and passengers to reflect on their most authentic feelings and engage in their most personal interactions due to its unique combination of privacy and exposure, intimacy and scope for rapid observation of the world. It is the place where boys and girls are told the facts of life. It is where first and last kisses are planted, where confessions are made, where secrets, money and drugs are exchanged, and where murders are committed, as in the *film noir* of the 1940s and '50s.

In the Hollywood *film noir*, the car is a death trap. As James Wolcott observed,

> no one in these B movies ever slides behind the wheel humming a happy tune, eager to enjoy a summer picnic or a Sunday outing . . . [Rather] escape is fuelled by fear and pursued by the furies . . . the atmosphere inside a film-noir getaway car is a foggy compound of cigarette smoke, cheap perfume, cold sweat, festering thoughts and the clinging past.[53]

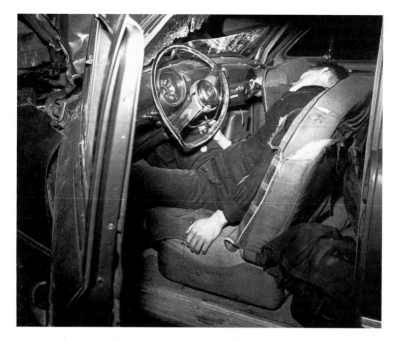

Many elements of automobile interiors proved lethal even in relatively minor accidents. The steering column killed many drivers, while passengers riding 'shotgun' in the 'death seat' were often thrown to their fate through unlaminated windscreens. Seatbelts became widely used only from the 1970s and airbags in the 1990s. The photographer Mell Kilpatrick carefully recorded such gruesome scenes for a Los Angeles insurance company in the 1950s.

The shadowy interiors, with padded ceilings and thickly framed windows of the heavily rounded vehicles featured in films such as *The Big Sleep* (1946) and *Kiss Me Deadly* (1955), are the sites of lies, betrayal and homicide. They were also used as coffins, as in *The Maltese Falcon* (1941) and, later, *Psycho* (1960).

The artist Edward Kienholz addressed the murky interior of the heavy, pre-war car in his assemblage of 1964, *Back Seat Dodge '38*. This construction, of chicken wire, artificial grass, plaster, a 1938 Dodge sedan and a couple of deformed dummies in coital position, presents the car's back seat as a grotesque hut, not a pleasure pit. Matthew Barney develops the theme of the metamorphic car interior in his *Cremaster* movies and sculptures resembling the human reproductive system. Andy Warhol's *Death and Disaster* silkscreen prints explored the darkest potential of the car as a dangerous space where real people are crushed within their living rooms on wheels. A decade earlier the photojournalist Mell Kilpatrick made a career recording the gruesome aftermath of real crashes, publishing photos that surveyed the details of wrecked car interiors: roofs collapsed, steering wheels mangled, seats upturned and doors torn off, the bodies of the fatalities still slumped or pinned in the wrecks.[54] Jean-Luc Godard's film *Weekend* (1967), Jim Dine's happening *Car Crash* (1960) and J. G. Ballard's novel *Crash* (1973) are among many other artists' responses to the interaction of human flesh with the distorted metal, smashed glass and shredded upholstery of wrecked cars.

The sheer thrill of speed, enhanced by creature comforts, such as radio, and by the display of form, colour and ornament that provide an expressive setting for the egos of drivers and passengers, long overshadowed the concern for safety in automobile design. While pointed bumpers and knife-sharp tailfins, such as those designed in Virgil Exner's 'stiletto studio' at Chrysler Corporation in the 1950s, were eventually recognized as dangerous elements on the outsides of cars, stylish bullet-like steering columns and other sharp protuberances on the interiors also contributed directly to countless injuries and deaths during the past hundred years. Novelties like Chrysler's Swivel Seats, offered on their 1959 line of cars, were touted to ease entry and exit from the low-slung models of the time. But they also threatened the ankles of rear-seat passengers and could eject their unbelted occupants overboard in a crash.[55]

Ford attempted to promote safety to the American public in the interiors of their 1956 models designed by William Burnett, yet the cars' seat belts, 'deep dish' steering wheels, padded dashboards and laminated safety glass in all windows left potential buyers unimpressed. The uneducated and brainwashed consumers of the 1950s wanted more gadgets, jazzier decor and quicker acceleration – never mind the brakes!

Despite such isolated efforts, until the 1970s design did not help much to address the alarming rise in accident statistics. But there had been other exceptions. Between 1927 and 1933 the architect-inventor Buckminster Fuller developed three versions of a streamlined automobile, the Dymaxion Car, which used the latest aircraft technology in the construction of its body and introduced a variety of radical design features in the interests of safety and economy. This was a large, three-wheeled, rear-engine car that synthesized characteristics of yachts, aircraft and contemporary abstract sculpture in an elegant teardrop design involving the talents of the aeronautical and naval designer Starling Burgess and the sculptor Isamu Noguchi. Powered by a Ford v8 engine that could take the big car to over 100 miles per hour, the vehicle boasted unprecedented manoeuvrability at low speed and also achieved high fuel economy thanks to its lightness and aerodynamics.

In the 1940s Preston Tucker attempted to produce a new post-war car that employed many innovative engineering and safety features, some of them, such as its rear engine, influenced by the Dymaxion car and the Czech Tatra. The Tucker Torpedo was billed as 'the safest, most aerodynamic car in the world' when it was unveiled in 1948. It boasted 67 safety features, including a perimeter chassis, pop-out windshield, four-wheel independent suspension, shock-insulated steering wheel, padded dash and pivoting third headlight to improve night-time vision.

Fifty-five Tuckers were built in a massive demobilized war production plant and a major advertising campaign was launched before the company was forced to liquidate, some say because of Detroit's hostility to anyone who challenged the status quo in car design. Nevertheless, a few crusading idealists, such as the lawyer Ralph Nader, whose book *Unsafe at Any Speed* (1965) revealed the callous design priorities of Detroit's Big Three manufacturers in the early 1960s, publicized the need for safer passenger cars and provoked Congressional hearings into the design practices of the Detroit manufacturers.

A vehicle never intended for production, constructed in 1958 as an experiment outside the North American automobile industry, was to point the way towards a safer car, particularly through the design of its interior. Built by the Cornell University Aeronautical Laboratory and the Liberty Insurance Company of Boston, the Cornell-Liberty Safety Car was called an 'ideas vehicle', a $250,000 laboratory on wheels, conceived to publicize car safety. It had no engine, since it was used exclusively for display, but it boasted a range of passive safety innovations, including anti-burst accordion doors, inertia-reel seat belts, an impact-absorbing frame and a fully upholstered interior with all potentially dangerous protuberances eliminated. The vinyl-trimmed passenger compartment accommodated six people in individual 'wrap-around' seats. The driver sat in the centre of the front row with passenger seats set slightly back to

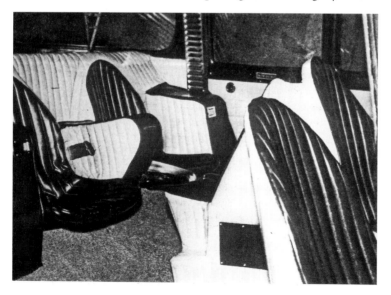

Experimental safety cars of the 1950s, produced outside the automobile industry, publicized interiors in which passenger protection was the highest design priority. The Cornell-Liberty Safety Car was the result of a collaboration between Cornell University and the Liberty Insurance Company. Among major manufacturers, the Swedish companies Volvo and Saab contributed most to the development of safer car interiors.

the right and left, so as not to obstruct the side views.[56] In the rear, the configuration was similar, but the central seat faced backwards. All accessories, such as door handles and window controls, were recessed and every surface was thickly padded. High seat backs were provided to reduce the danger of whiplash injuries, commonly resulting from rear-end collisions. Although the c-l Safety Car was mercilessly ugly, many of its cabin design features were eventually incorporated into the products of the major manufacturers. And the idea it promoted, of the safer interior, became a more significant aim of mainstream car designers from the 1970s onwards.

Some significant safety innovations were devised for the cockpits of racing cars to protect both their control mechanisms and the driver. The cockpit, known in racing circles as 'the tub', placed at the centre of the car, is its operational centre and the safety cell for the driver. It provides the structural core of the chassis with the engine and front suspension clipped to it. In recent examples, the cockpit is typically built of strong carbon fibre panels with honeycomb and woven elements for impact absorption. The tub is also a bag of tricks, including roll-over bar, crumple zones, rapid-release steering wheel for quick escape and elasticized helmet restraints to prevent serious whiplash injuries. Although the materials and construction technology of racing cars are too expensive or inappropriate for passenger cars, the ideas they generate have often been transferable.

While the American response to safety design initiatives was initially feeble or obstructive, in Europe some manufacturers made greater efforts to design safety into their new models. The British Rover 2000 of 1962 featured a rigid passenger cell and crumple zones front and rear to absorb the energy of a collision. It also included a collapsible steering column, child-proof locks on the rear doors and a dashboard made of honeycomb, impact-absorbing plastic and urethane foams of varying densities. Italy's leading *carrozzeria*, Pininfarina, designed and built the experimental Sigma PF in 1963 to demonstrate that a safe car could also be attractive. The Sigma's sleekly rounded body featured sliding doors, anti-roll bars in the roof, extensively padded interior surfaces and ergonomically contoured seats, constructed like those of a racing car to provide driver and passengers extra protection in a collision.

The most influential work, however, was done in Scandinavia. Spurred on by the example of Swedish manufacturers, Saab and Volvo, car makers around the world gradually became more focused on the improvement of their cars' crash avoidance and resistance capabilities, interior design and eventually even pedestrian safety. The Swedish makers were the first to introduce the three-point seat belt, designed for Volvo by Nils Bohlin in 1959, and they developed or were the first to use a passenger safety cage, effective crumple zones, impact-absorbing bumpers, collapsible steering wheel, rear-facing child seat, head restraints, multiple airbags and inflatable side-impact curtains. Saab, whose company foundations were in aeronautics, employed Gunnar Ljungstrom, Sixten Sason and Bjorn Envall for engineering and design with a slant on safety. Over the decades, Swedish cars earned a reputation for above-average crash performance due to their many innovations since the 1940s. Their popularity around the world forced, often reluctant, more style-conscious competitors to privilege safety in the design process. They also demonstrated that design reform *can* precede regulation.

## Benches, Beds and Buckets

An automobile's appearance had been important since the start of motor manufacturing, but the campaign to commercialize style began in earnest with the formation by General Motors in 1927 of their Art and Color Section, headed by the flamboyant Californian Harley Earl. Earl and the President of General Motors, Alfred Sloan, were dedicated to converting the car body into a mannequin, which they would dress in an annually changing procession of the latest ornaments, gadgets, paints and fabrics. These would be arranged in different ways each year to maintain an image of freshness about the product and to conceal the slow evolution of the car's mechanics. In the early 1930s styling also spread to companies that were known for their engineering bias. Art Deco-style hardware and upholstery patterns were applied to soften the innovative, functionalist appearance of Carl Breer's Chrysler Airflow. The Airflow's interior used hygienic marble-patterned rubber mats instead of carpeting on the floor, and its seat cushions were cradled in exposed tubular steel frames resembling the steel furniture being produced at the time for domestic and office use by various manufacturers exhibiting alongside Chrysler at the Century Progress Exposition in Chicago (1933–4). Typical among them were stylish designs by Donald Deskey, mass-produced by the Ypsilanti Reed Furniture Company, and those by Gilbert Rhode for Herman Miller.

The evolution of the car seat, from a thinly padded wooden bench to the electronically personalized driver's chair available in today's automo-

The visible tubular steel furniture frame became a signature of advanced interior design in the 1930s. The chrome-framed and substantially cushioned sofas inside the Chrysler Airflow would have looked equally appropriate in the lobby of Radio City Music Hall.

The driver and front passenger seats aboard the 1906 Brasier bore a close resemblance to the substantial furniture that the car's Edwardian occupants would have sat in at home or in their offices. Curvaceous forms, thick padding and opulent buttoned leather upholstery provided compensation for the car's otherwise cold comforts.

biles, has come to lead other types of furniture in innovative design and construction. Car seats have, however, also evolved in relation to popular taste in furniture design. The more elaborate of Edwardian cars, Mercedes or Rolls-Royce, featured highly sculptural seating with contoured backs that wrapped around the occupants, like the curvaceous sofas of the period. Upholstery too was designed in line with conventional domestic taste, employing deep tufting and buttoning, with leather covering to resist the elements in what were mainly open-topped vehicles.

By the late 1920s, when the closed car had become more common, broadcloth and leatherette were the standard upholstery coverings. The full-width bench seat, movable forward and back on tracks to accommodate drivers of different arm and leg lengths, was the first concession to what would later be called ergonomics. Bench seats accommodated two or three people across, depending on the width of the car body. Yet the fixed relationship between the planes of seat cushion and back (squab) limited access to the rear seat of two-door models until the early 1930s, when the squab was divided in two sections, each hinged to fold forwards independently. This not only eased entry to the back seats of coupés and convertibles, but also allowed them to be built with narrower doors that were more convenient in parking lots and garages.

In the late 1930s, when motels were still a rarity in many places, the potential of the car as overnight accommodation was exploited most aggressively by the Nash company, who patented a reclining seat that offered motorists the flexibility to convert the interiors of their cars into a full-length double- or triple-width bed. Advertisements often illustrated these in a night-time configuration, made up with sheets and blankets like a Pullman sleeper berth or a Berlin travelling carriage. Nash continued to promote their convertible interiors into the 1960s, when camper vans, motor homes and caravans were widely available as alternatives to motels and motor courts, which were by then more common roadside features in developed countries around the world. The attractions of a fully reclining seat, however, were lost neither on other manufacturers nor on those motorists whose idea of a bedroom on wheels harks back to Casanova's post chaise.

In the mid-1950s European manufacturers such as Lancia, Facel-Vega and BMW were building large luxury sport coupés, described as grand tourers, that had evolved a 2+2 seating configuration. In these cars the gearshift and handbrake were located above the transmission tunnel, flanked by individual 'bucket seats', front and back. Americans soon discovered the pleasures of these stylish high-performance cars, and to tap this market niche Detroit manufacturers began to develop new products. The four-passenger Ford Thunderbird of 1958 was the popular pioneer of this type, a reasonably priced alternative to the often temperamental and expensive European thoroughbreds.

While the exterior was extravagantly styled under the direction of George Walker and Joseph Oros, the t-Bird's most influential feature was its sporty seating plan. Deeply sculptured panels of the bodywork flowed inside to form a twin-pod dashboard from which a full-length transmission console swept down to divide the front bucket seats and the distinctively contoured rear bench. Radio controls and speaker, heating controls, ashtrays and electric window switches were mounted in the transmission console, turning this intrusive mechanical element into a trend-setting feature.[57] Seats were covered in a combination of Haircell-grain vinyl with insets of woven nylon seat cloth or leather. The individuality of the seating plan was contrived specifically to distinguish this 'personal' car from the standard six-passenger family sedan.

Despite such typological innovations as the Thunderbird, in larger family cars on both sides of the Atlantic during the mid-1950s, such as the French Panhard Dyna z, the standard automobile seat remained a sofa accommodating three passengers side-by-side, front and back. These seats consisted of a tubular steel framework, coil springs, covered with kapok or other thick wadding, and tufted, pleated or buttoned upholstery, colour coordinated with headlining, door panels and carpets. New synthetic upholstery materials including nylon and leather substitutes, such as Naugahyde, were manufactured in bright pastel colours and arranged in exotic combinations to complement the multi-hued exterior paintwork fashionable at the time. Such new materials enhanced the annually changing fashion show of seasonal colours and patterns within a relatively unchanging internal plan.

Textile designers such as Marianne Strengell and Jack Lenor Larsen and the weaver Gerda Nyberg, all associated with the Cranbrook Academy of Art outside Detroit, designed upholstery materials specifically for the automobile industry. Their work translated the unique qualities of hand-woven textiles into mass-produced fabrics using the most advanced synthetic fibres. In addition to rich materials, upholstery piping and chromium trim were used extensively to outline the dramatic contours of the seats, the sculpted interior door panels, armrests, dashboard, and the pile or twist carpeting that often

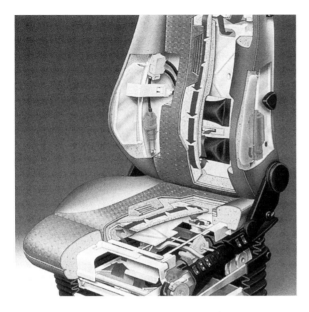

Coach and lorry drivers are assisted in their work by highly sophisticated seats. The Recaro Ergo Metro professional driver's seat incorporates a variety of technologies to provide its occupant with highly tailored and fully adjustable support. The seats employ separately adjustable air chambers, internal heating and variable density foams to provide comfort and reduce chronic back problem suffered by many professional drivers.

spread up from the floor on to the door sills and transmission tunnel.

Adjustable or integral headrests, added for both safety and comfort first by Saab and Volvo and universally in the 1970s, also became a prominent visual feature of the car seat, altering its proportions and approximating the form of a domestic lounge chair. Elaborate motorized or manual gear-operated mechanisms controlling height, rake and horizontal position further enhanced the ergonomic comfort of the driver's seat from the early 1950s to present-day models, which can be programmed to remember multiple drivers' preferences and measurements. Recent Cadillac automobiles have been fitted with BCAM International's innovative 'intelligent' passenger car seat, which houses a computer in the squab to control the contour of the entire seat surface, automatically custom-fitting any occupant. This dynamic system provides specific lumbar, thigh and back support, while frequently checking the comfort of the occupant and readjusting if necessary.[58]

As early as 1940, the industrial designer Walter Dorwin Teague wrote: 'Automobile manufacturers have made, in the past few years, a greater contribution to the art of comfortable seating than chair builders in all preceding history.'[59] With the incorporation of movable armrests, built-in heating elements, integrated audio speakers and even massage functions, the modern car seat has progressed significantly beyond normal domestic (or even office) seating in its technology and the comfort it can offer.

Ultramatic, the compelling name given to the automatic transmission system used by Packard after the Second World War, could equally describe all the sophisticated mechanical advances designed to make the driver's life as simple and undemanding as possible. While today car manufacturers around the world provide such mechanical comforts, they were exploited first in North America in order to get more people driving. Automatic transmission was offered for the first time on a mass-produced car, General Motors' Oldsmobile, in 1940. In the same year electric window lifts were standard equipment on the Lincoln Continental.

Power steering, power brakes, electrically adjustable seats, automatic headlight dimming, station-seeking hi-fidelity radios, electrically operated convertible roofs and electric radio aerials were all commonly available on American cars by the early 1950s. By then, the great majority of North Americans had a driver's licence.

Also in that decade, the interiors of General Motors' Detroit cars were 'feminized' by the first team of women to be employed by the design department of a major automobile manufacturer. The women were assigned to a special section of the interior design department, where their work was limited to the selection of colours, finishes and accessories, intended to seduce both male and female buyers. Their innovations included glove compartments fitted with vanity mirrors and make-up holders, fur-look upholstery trimming and an emphasis on dramatic colour combinations – hot pink with black was a favourite duo – while publications about the cars were like fashion spreads or home decoration advertisements. Today, the status of women in automotive design remains marginal and attracts special notice when they have the rare opportunity to design something significant.[60]

The simplification of driving and the feminization of car design in North America were both products of economic and political imperatives in the period from 1930 to the 1970s. Highway construction was a major element of the New Deal recovery programme of Franklin Roosevelt's presidency during the Great Depression of the 1930s. Encouraging women to drive and to buy cars was seen as an essential aspect of economic recovery, and the two major elements of the programme were the automatic car and the superhighway (or motorway). The design of a national motorway network with wide traffic lanes, smooth surfacing, clear directional signage, elimination of traffic lights, one-way entry and exit ramps, and bright overhead lighting after dark made driving easy and pleasant for even the most timid of motorists. The construction of such major engineering projects also provided jobs as part of the great public works programmes of the post-New Deal era.

The military advantages of a modern motorway network were also apparent to the governments that pioneered their construction, both in North America and in Europe. The leaders of the German Third Reich saw the potential for rapid troop movements over an autobahn network and supported the development of passenger cars specifically for use on it. The KDF Wagen (Volkswagen), designed by Ferdinand Porsche in the mid-1930s, was a direct response to this perceived need, although it was not produced in large numbers until after the Second World War. Its streamlined shape and air-cooled rear engine were designed for travel over the smooth surfaces of the autobahn at continuous high speeds, while its roomy interior for four offered good heating and ventilation for passenger comfort, excellent all-round visibility, light steering and a simple transmission for ease of handling.

Motorways have now become a ubiquitous feature of both urban and rural landscapes around the world, and their impact on the natural and human environments is the subject of fierce debate today. Yet, there is little argument that they comprise one of the most significant features of the modern physical world, and they provide the primary infrastructure for our mobile lives. While air traffic patterns and shipping lanes are practically invisible, rail tracks and motorways form a monumental part of the built environment. From the perspective they offer us, in our cars and trains, we see and understand the rest of the physical world in a specifically modern way.

## One for the Road

The extremes of style employed in motorcar interiors were perhaps most evident in the work of custom coachbuilders active in the early years of the twentieth century, when extravagant, individually designed, chauffeured limousines were still a status symbol in a league with private Pullman cars and custom-built yachts. Typical among them is a Rolls-Royce Phantom, built for the Woolworths director C. W. Gasque in 1927 by the small English coachbuilders Clarke & Son of Wolverhampton.

The builders had an unlimited budget to turn the interior of a stately, but unremarkable town car into a Rococo drawing room, which was intended as a romantic gift for Mrs Gasque. The ceiling was painted with cupids and pink roses; wall panelling was decorated with arabesques and medallions depicting *amoretti*; and its Louis xv sofa was upholstered in rich Aubusson tapestry. Interior lights in the form of torches were held aloft by gilded cherubs, as was the French ormolu clock placed at the centre of a bow-fronted Neo-classical cabinet containing the cocktail bar, from which the Gasques could take one for the road.

The car's rectilinear body allowed for the creation of a passenger compartment that functioned as a fully glazed observation room, in which each of the perfectly rectangular, gold-framed windows created a distinct pictorial view of the world outside. This pastiche eighteenth-century French interior was a scaled-down version of the sort of rooms that William Randolph Hearst was commissioning for his California palace, San Simeon, at the same time.[61] Since the dawn of motoring, such extravagant interiors have been fitted to Rolls-Royce, Cadillac or Mercedes chassis in many styles to satisfy the diverse tastes of popes, maharajas and 'five and dime' or 'dot-com' tycoons.

Although styles have changed and the market has broadened, extravagant custom-built automobiles are still being commissioned with bodies and interiors that reflect the particular interests of the owner. Beginning in 1950s California, the popular desire for unique automobiles was satisfied by the custom-car movement, which sprang up in Los Angeles and spread across the country and abroad in the following decades. Professional customizers, such as George Barris, and amateur craftsmen transformed standard products of the assembly line into special works of machine art. Extravagantly altered body shapes and flamboyant paint jobs were accompanied by fanciful interiors in which the major ingredients were typically tuck-and-roll or tufted-and-buttoned Naugahyde upholstery over specially designed bucket seats, carpets you could lose a shoe in, fur-trimmed dashboards, elaborately upholstered headliners and the ubiquitous cocktail bar. Special high-performance sound systems were also important additions, as were the signifiers of a car's theme, also reflected in its name: *Miss Elegance, The Bat, Lost in the fifties, Heartbreak*.

Since such cars were owner driven, and the instrument panel connected the driver with the immense power of the souped-up engine, dashboards became a focus of the customizers' attention. There, the maximum amount of equipment was desirable and the drama of its presentation essential. Banks of toggle switches, often with ornamental knobs attached, large areas of chromium plating, sculptural shift levers and hand-brake controls, multi-coloured, leather-wrapped and accessorized steering wheels, and symbolic embellishments, such as furry dice and miniature religious statuary, all added personal character to the interior of the custom car. Above all, the distinguishing feature of these interiors was colour, which became more expressive in the 1950s due to the development, by Dupont and other chemical companies, of vividly coloured vinyl fabrics, nylon carpeting and metal-flake synthetic lacquers.

From the 1950s, the design of production-car dashboards also became an exploration of weird and fanciful anthropomorphic shapes, intended to divert and amuse driver and passengers with the equipment it contained and the form it took. The 1961 Plymouth demonstrated the science-fiction influence evident in the control panels of US cars built

Customizers transform the most mundane and common mass-produced cars into unique works of popular art by sculpting body metal and by composing a diverse range of stock components. Their interiors reflect the most personal of their owners' interests through informed design choices and high craft skills.

The French-built Smart car features an anthropomorphic dashboard that contributes to the car's benignly hi-tech character – the friendly personal robot.

in the era of UFOS and the space race. The Plymouth employed a face-like instrument cluster containing a horizontal, bar-shaped speedometer joining two large circular pods, the three elements linked by a projecting monobrow shading the two circular 'eyes' and stretching across the length of the speedometer. Below them, set into a broad, bulging 'chin' was a rectangular 'mouth' containing temperature, battery, oil gauges and the centrally mounted clock. A square, gold-flecked, transparent Lucite steering wheel framed the bizarre cluster of instruments, giving the configuration the appearance of a startled alien from an *Adventure Comix* magazine.

Designs for Detroit cars of mid-century were both literal and abstract and open to a variety of readings. The body of the 1958 Cadillac combined jet-age symbols with distinctly anthropomorphic elements in the bodywork. Eyes, breasts, mouth and the more ambiguous curves suggesting shoulders and hips were treated in a hard, geometrical and mechanistic manner. Their ambiguity and complexity of meaning related to the tradition of robots or automata. Yet, the hardness of the metal, its razor-sharp edges trimmed with chromium for emphasis, was softened by subtle, muscular swellings of fender and door surfaces and by the luxurious tactile qualities of the interior. There, richly coloured carpets, brocade upholstery and soft leather produced a lush, enveloping environment, which isolated the occupants from the external world. The driver was physically lulled in the fully automated and climate-controlled cabin, but the senses were continually stimulated. A muffled burble from the massive V8 engine, multiple-speaker high-fidelity radio, whirring of motorized appliances, light glinting on chromium-plated ornaments and emanating mysteriously from dashboard instruments, all kept eyes and ears working in a pleasurable symbiosis with the machine.[62]

Only French manufacturers, particularly Citroën, matched or exceeded the theatricality of American mass-produced cars from the 1930s to the 1970s. Today, those same excesses are, to some extent, echoed in the electronic drama of satellite-navigation screens and the digital display of iconic images in the instrument panels of upscale German and Japanese cars such as Audi and Lexus. The dashboards of recent cars have taken a number of divergent directions, retaining traditional elements, becoming more ergonomically correct and predictably dull, or anticipating a major change through the introduction of new technologies.

The collective memory of coach-building traditions is kept alive in some executive-class cars of the early twenty-first century. For those who appreciate the classic symbols of tradition, the form of many gauges in today's 'retrofuturist' cars harks back to the earliest dashboards, which were wooden panels inset with round instruments, which emulated the form of the clock.[63] British car makers, such as Rover, Bentley and Jaguar, have traditionally finished their interiors with leather and walnut. These have long remained crucial signifiers of privilege and of enduring high standards of craftsmanship, despite the foreign ownership of all three companies and the globalization of the cars' mechanical and structural elements.

At any given time in the history of motoring, there has been a dominant feature in the typical dashboard. At first, it was the speedometer and the clock, since time and speed were the raisons d'être for the car. By the 1930s the radio had come to the centre of the dash, as it had to the living rooms of most homes. With the fuller expression of styling from the 1950s, designers located elaborately sculpted clusters of instruments and controls directly in front of the driver. The rest of the panel housed the glove box, the radio and heater controls, some of the latter combined with bulky air-conditioning

Satellite navigational systems, installed in vehicles of all types, have facilitated and expanded the process and effectiveness of map reading. They have assumed the dominant position in the automobile dashboard.

units, which represented the ultimate in sybaritic motoring along Miami Beach or the Côte d'Azur.

Now, the VDU screen holds the key to the management of a highly intelligent, computer-controlled interface between the driver and the machine, including satellite-navigation showing the driver where he or she is going and the energy management systems of a new generation of hybrid cars, such as the Toyota Prius. As we move into the new century, designers predict that the most typical dashboard will come to resemble the glass cockpit of current commercial airliners, with a fully electronic display, giving the driver only the information requested by voice command. It is also likely that visual information will be presented in the car's windscreen, as in the helmet visors of military pilots.

Recent concept models suggest that the layout of car interiors will become more diverse and will cater more directly to the differing lifestyles of their individual owners. The family car, which has grown steadily larger in the 1980s and '90s, becoming a multi-purpose people mover, such as the Renault Espace, is predicted to become even more of a home on wheels, with areas devoted specifically to children's activities, with more flexible seating space, better provision for eating on the move, and for enjoying all the potentials of Internet, telephone and television. 'One for the road' may refer no longer to a Daiquiri, but to the ubiquitous Starbucks' coffee in a retractable cup holder next to every seat.

The future of car design – wild or mild? Marc Newson's stylish concept design for Ford (above) retains traditional elements of car space, while the Toyota Pod (below) presents the automobile cabin as a multi-functional and media-rich cyber-café on wheels.

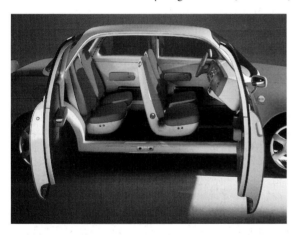

Retro-styled vehicles, such as the Volkswagen Beetle, the Mini Cooper and the Mazda MX5, have developed a new stylistic niche, giving their ageing baby-boomer drivers a cosy sense of continuity with the past while enjoying the advantages of the latest automotive technology. The Beetle, for example, incorporates a bud vase in its dashboard, reminiscent of early electric-powered ladies' cars and of vw's association with 1960s flower power.

Despite the increasing difficulty of car use in urban areas, rises in the cost of fuel and environmental damage, sport utility vehicles are thriving in terms of popularity and interior design innovation, combining a sense of massive defensiveness with extreme sport capabilities. The ultimate vehicle in this category is the Hummer, a serious battle-wagon, redolent of US adventuring in the Middle East, tamed through the luxurious finish of its interior, adapted for the daily school run or presented as a linear cocktail lounge in the stretched limousine versions.

Many architects and designers have contributed to the development of new concepts in automotive design. The collaboration between J. Mays and the multi-disciplinary designer Marc Newson on the design of the Ford Concept 021C in 1999 provided an interesting glimpse of what the near future could hold. The return of the swivel seat, this time in a safer and more practical form, is the most striking idea coming from the

The erotic potential of car seat design was fully exploited by General Motors in their Chevrolet Monte Carlo of 1974, and emphasized, without the benefit of subtlety, in this promotional photograph.

furniture designer's perspective. With its two-tone paintwork, analogue instruments in round, chrome-plated bezels and centre-opening doors, which hark back to elegant Ford sedans of the past, Newson's 021C combines classic and contemporary elements while avoiding the retro-futurist cliché of directly imitating well-loved models from the past.

Architects have also turned their hands to car design. Le Corbusier's radical Automaxima of 1928 was a streamlined people's car with banquette seating for three in front and storage space behind. Less radically, Walter Gropius produced several conventionally elegant designs for Adler of Germany between 1927 and 1931, pioneering the use of reclining seat backs. After completing a radically innovative factory for BMW in Leipzig in 2004, Zaha Hadid also turned her hand to concept car design, creating an asymmetrical, three-wheel pod vehicle, the z-Car. Intended to be powered by a hydrogen fuel cell, the two-passenger z-Car features a unique hinged rear suspension that would enable the car to change its height and wheelbase for either city driving or highway cruising. Inside, its organically formed steering wheel would contain all the car's 'drive-by-wire' controls, leaving the rest of the interior completely clear of switches or levers. This prototype is as close as car design has come to the Deconstructivist architecture of the late twentieth and early twenty-first century. The Italian architect Mario Bellini designed the fantastic Kar-a-Sutra, a lime-green, open-topped, fibreglass van, its interior furnished with foam cushions for a sensuous, communal ride. Although it was never produced, the prototype became a touchstone for freewheeling sexuality in the early 1970s.

If, in the past, it was commonly thought that travel by train and boat could lead to moral degeneration through drink and generally reckless behaviour, they have been nothing as compared with the influence of the automobile on the development of personal and sexual freedom in its hundred-year history. In his 1945 novel, *Cannery Row*, John Steinbeck wrote of the Model T Ford's role in the sexual liberation of the American nation: 'Most of the babies of the period were conceived in Model T Fords and not a few were born in them.'[64] Later, in the aggressive sexual climate of the 1960s, another specific correlation, between a particular make of car and its ability to arouse, was demonstrated in Russ Meyer's cult horror film *Beyond the Valley of the Dolls* (1970), in which one of Meyer's predatory temptresses entices her lover into the walnut-panelled and leather-upholstered rear seat of a Rolls-Royce for limousine sex. As her climax builds, she chants ecstatically: 'There's nothing like a Rolls, a Rolls, a Rolls! (gasp) It's better than a Bentley! (gasp) There's nothing like a Rolls, a Rolls, a Rolls, (screaming) A ROLLS!'

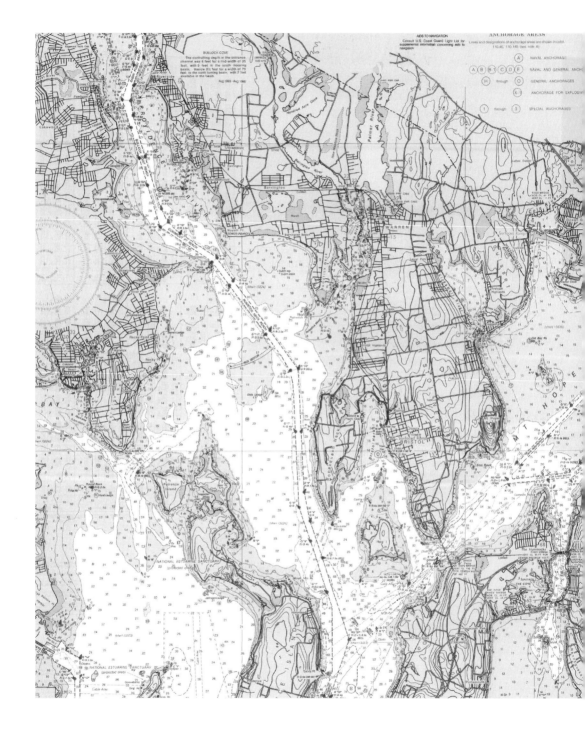

# 2 WATER

## Smoke and Mirrors

> I shall never forget the one-fourth serious and three-fourths comical
> astonishment, with which, on the morning of the third of January
> eighteen-hundred-and-forty-two, I opened the door of, and put my
> head into, a 'state-room' on board the Britannia steam-packet . . .
> bound for Halifax and Boston, and carrying Her Majesty's mails.

Thus 'Charles Dickens Esquire and Lady' began a voyage to America that
would inform both his published memoirs of the trip, *American Notes*
(1842), and his novel *Martin Chuzzlewit* (1843–4). Dickens's ironic
account of his cramped 'state-room', the grim public spaces and the general discomfort of an eighteen-day transatlantic voyage on one of the
world's first steam-powered, purpose-built ocean liners making regular
scheduled crossings between England and North America provides very
graphic insights to aspects of early modern travel still with us today.

Built for Samuel Cunard in 1840, *Britannia* was the first steamship
commissioned specifically for the North Atlantic mail and passenger
trade. The ship looked every inch a clipper, with three masts and a full
complement of sails. Her tall smokestack and gigantic side-mounted
paddle wheels, however, told the world that this was something new – a
ship that could cross the ocean on schedule, regardless of the weather, in
as little as half the time of her sail-only contemporaries. *Britannia* carried up to 115 passengers in accommodations that were trumpeted to be
luxurious enough to please her wealthy and adventurous clientele. Thus,
Dickens's disappointment on first sight of his stateroom:

> this utterly impracticable, thoroughly hopeless, and profoundly pre-
> posterous box, had [not] the remotest reference to, or connection
> with, those chaste and pretty, not to say gorgeous little bowers,
> sketched by a masterly hand, in the highly varnished lithographic
> plan hanging up in the agent's counting-house in the city of
> London.

Sea routes through
Narrangansett Bay.

99

As is often the case today, the reality of travel accom-modations then did not live up to the hyperbole and pictures in the travel agent's office. Dickens's tiny pan-elled cabin contained two narrow bunk beds, with luggage space under the lower one, an upholstered 'shelf' on which one could sit, a small cabinet with pitcher and bowl, a mirror and a porthole. The writer suggested that his introduction to the ship might be a joke, 'a pleasant fiction and cheerful jest of the cap-tain's, invented and put in practice for the better relish and enjoyment of the real state-room presently to be disclosed'. However, he asserted that in reality 'nothing smaller for sleeping in was ever made except coffins'.

Dickens's disappointment in the accommodation foreshadowed his disappointment with America dur-ing the four months he spent there. And in terms of the ship, his bemusement was not reserved for his cabin. The main saloon of the ship also failed to match the expectations created by Cunard's advertising, which showed:

> a chamber of almost interminable perspective, furnished . . . in a style of more than Eastern splen-dour, and filled (but not inconveniently so) with groups of ladies and gentlemen, in the very highest state of enjoyment and vivacity.

The reality, according to Dickens, was

> a long narrow apartment, not unlike a gigantic hearse with windows in the sides; having at the upper end a melancholy stove, at which three or four chilly stewards were warming their hands; while on either side, extending down its whole dreary length, was a long, long table, over each of which a rack, fixed to the low roof, and stuck full of drinking-glasses and cruet-stands, hinted dismally at rolling seas and heavy weather.

He noted the importance on board *Britannia* of personal service, which would become the hallmark of all de-luxe travel in the future: 'God bless that stewardess for her piously fraudulent account of January voyages!' He also confirmed that grand dining was already an established feature of life aboard a North Atlantic passenger steamship:

Isambard Kingdom Brunel's *Great Britain*, launched in 1842, is very similar to the SS *Britannia*, on which Charles Dickens first crossed the Atlantic. The spartan cabin and public rooms disappointed the writer, yet such ships embodied many of the basic design characteristics associated with all later liners.

Turtle, and cold Punch, with Hock, Champagne, and Claret, and all the slight et cetera usually included in an unlimited order for a good dinner – the dinner of that day was undeniably perfect; that it comprehended all these items, and a great many more; and that we all did ample justice to it.

Conviviality among the passengers was a matter of urgency. Dickens reported that everyone on board felt compelled to make the most of their fellow passengers' company and to avoid any conversation that would disturb the equilibrium of the more nervous travellers among them. As the weather worsened, they tolerated each other's dishevelment and nausea, and they attempted to provide mutual support and comfort over the long, rough haul between Liverpool and Boston. Describing the physical discomfort of a midwinter crossing, Dickens pointed out the frightening appearance of the sea and sky and the effect that the dramatic weather had on the shipboard community:

> the sea ran high, and the horizon encompassed us like a large black hoop. Viewed from the air, or some tall bluff on shore, it would have been imposing and stupendous, no doubt; but seen from the wet and rolling decks, it only impressed one giddily and painfully. In the gale of last night the life-boat had been crushed by one blow of the sea like a walnut-shell; and there it hung dangling in the air: a mere faggot of crazy boards. The planking of the paddle-boxes had been torn sheer away. The wheels were exposed and bare; and they whirled and dashed their spray about the decks at random. Chimney, white with crusted salt; topmasts struck; storm-sails set; rigging all knotted, tangled, wet, and drooping: a gloomier picture it would be hard to look upon.

Thus, despite Dickens's charm and irony, the genuine hardships of the voyage were ever present in his narrative. He described the isolated culture that developed on board during the crossing:

> four good hands are ill, and have given in, dead beat. Several berths are full of water, and all the cabins are leaky. The ship's cook, secretly swigging damaged whiskey, has been found drunk; and has been played upon by the fire-engine until quite sober. All the stewards have fallen down-stairs at various dinner-times, and go about with plasters in various places . . . News! A dozen murders on shore would lack the interest of these slight incidents at sea.[1]

Any modern voyager who has travelled on a ferry in a rough sea will have some slight inkling of what Dickens went through on board *Britannia*. The sea is still a siren, beckoning the adventurer in many of us, yet it remains a potentially hostile environment. What follows is an account of the ways we have used design to help us survive and enjoy our journeys on the waterways of the world.

## On Rivers and Canals

Patent furniture, such as reclining chairs and convertible beds, that provided comfort for mid-nineteenth-century rail travellers, first in the US and later in Europe and around the world, began its development in trades such as wagon building and the outfitting of canal boats. Although little physical evidence remains of the interiors of the packet boats that plied the inland waterways of the early United States, personal accounts recall innovative interiors that were contrived to make the most of limited space in order to provide passengers with a comfortable voyage. The substantial length of these ships, up to approximately 100 feet, and their beam of around 14 feet created an axial interior space that provided a model for later railroad carriages. And to form a civilized saloon and dining room for daytime use and sleeping quarters for the nights, all in the same space, interior designs were based on the principle of convertibility.

A British traveller, Thomas Woodcock, writing in 1836, reported that the entire length of the boat on which he travelled the Clinton (aka Erie) Canal was a single space:

> the forward part, being the ladies Cabin, is separated by a curtain, but at mealtimes this obstruction is removed, and the table is set the whole length of the boat. The table is supplied with everything that is necessary and of the best quality with many of the luxuries of life.

Woodcock went on to marvel at the adaptability of this single space to the problem of sleeping on board:

> the Yankees ever awake to contrivances have managed to stow more in so small a space than I thought them capable of doing. The way they proceed is as follows – The settees that go the whole length of the boat on each side unfold and form a cot bed. The space between this bed and the ceiling is so divided as to make room for two more . . . the space between berths being barely sufficient for a man to crawl in . . . The first night I tried an upper berth, but the air was so

foul that I found myself sick when I awoke. Afterwards I chose an under berth and found no ill effects from the air. These boats . . . go at a quicker rate and have the preference going through the locks, carry no freight, are built extremely light, have quite Genteel men for their Captains, and use silver plate.[2]

Contemporary accounts suggest an atmosphere of civility and conviviality on board these ships, with the travellers engaged in something like a pyjama party afloat. It is possible, however, to imagine that the level of human engagement on an extended voyage would not suit everyone. The American novelist Joseph Ingraham, travelling by river steamer during the 1830s, revealed his need for privacy among his fellow passengers:

Having secured a berth in one corner of the spacious cabin . . . I could draw the rich crimsoned curtains around me, and with book or pen pass the time somewhat removed from the bustle, and undisturbed by the constant passing of the restless passengers.[3]

Nathaniel Hawthorne also noted the reliance on curtains as spatial dividers when describing his experience aboard a packet boat in the mid-1840s:

[With] the crimson curtain being let down between the ladies and the gentlemen . . . the sexual division of the boat . . . the cabin became a bedchamber for twenty persons, who were laid on shelves one above another . . . Forgetting that my berth was hardly so wide as a coffin, I turned suddenly over, and fell like an avalanche on the floor.[4]

These primitive arrangements offered considerable opportunities for public embarrassment, yet the curtain would remain a significant form of spatial divider in tubular transport interiors such as Pullman cars and, later, in airliners.

If apparent privacy was to become one of the primary aims of transport interior design, pleasure was also drawn from many of the features of the early steamboats. An anonymous American, travelling on the Clinton Canal in 1829, described his voyage with lyrical enthusiasm: 'Never was I more delightfully situated as a traveller than on this occasion [as] our boat skimmed its peaceful way along this artificial and wonderful water communication.' The boat on which he travelled was a superior freight carrier that also took passengers in two cabins, fore and aft of the freight compartment, each about 23 feet long and sufficiently high for a six-footer to stand erect with his hat on. The roof is in the form of the back of a tortoise, and affords a handsome promenade, except when the everlasting bridges open their mouths for your head.

He went on to report the distressing case of a young Englishwoman, who had 'met with her death a short time since, she having fallen asleep with her head upon a box had her head crushed to pieces' as the boat passed under one of the canal's many low bridges.[5] Thus, even the gentlest forms of early transport presented physical dangers as well as amazing opportunities.

Larger riverboats with several decks and tall paired chimneys provided a much more elevated viewpoint from which to observe and discover the country than the low-built canal boats with their collapsible smokestacks. Anne Royall documented the sheer pleasure of seeing countryside, settlements and human activity from the novel perspective of a Mississippi River steamboat on which she travelled in the American South during the 1820s:

> The banks are lined and ornamented with elegant mansions, displaying in their richly adorned grounds, the wealth and taste of their possessors; while the river, now moving onward like a golden flood, reflecting the mellow rays of the setting sun, is full of life. Vessels of every size are gliding in all directions over its waveless bosom, while graceful skiffs dart merrily about like white-winged birds. Huge steamers are dashing and thundering by, leaving long trains of wreathing smoke in their rear. Carriages filled with ladies and attended by gallant horsemen enliven the smooth road along the Levee; while the green banks of the Levee itself are covered with gay promenaders. A glimpse through the trees now and then, as we move rapidly past the numerous villas, detects the piazzas, filled with the young, beautiful, and aged of the family, enjoying the rich beauty of the evening . . . the deep trombone of the steam pipe – the regular splash of the paddles – and the incessant rippling of the water eddying away astern, as our noble vessel flings it from her sides, no longer affect the senses, unless it may be to lull them into a repose . . . The shores, conse-

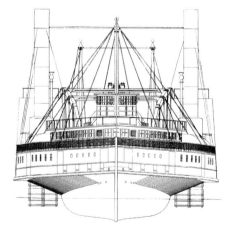

The Mississippi River steamer *Commonwealth* was a beamy, shallow-draft barge with a wooden hotel built on top. Yet such vessels were among the wonders of their age and revolutionized the culture of travel on inland waterways in North America.

North American river steamboats were genuine peoples' palaces that both reflected and influenced the nation's taste. Above is the Renaissance Revival-style saloon of the Hudson River steamer SS *Drew*, its rich architectural detailing enlivened by vibrantly coloured carpets, bright paintwork and, above all, gilt.

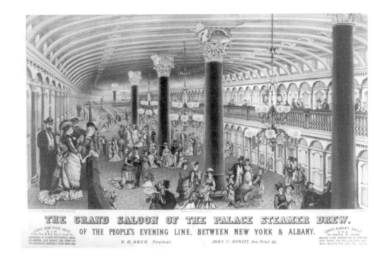

THE GRAND SALOON OF THE PALACE STEAMER DREW.
OF THE PEOPLE'S EVENING LINE, BETWEEN NEW YORK & ALBANY.
W. H. DREW, President. JOHN C. HEWITT, Gen. Ticket Agt.

quently, present from the lofty deck of a steamer . . . a very singular appearance.[6]

Royall's boat would have been similar to the *General Pike*, built in 1819, which set a standard of comfort and elegance for passenger river craft in the American west. *General Pike* was a wide, flat-bottomed barge with a grand hotel of several storeys built on top and decorated in the Greek Revival style, which was fashionable in the first half of the century. The captains of such ships, many of whom up to mid-century owned their boats, expressed their own taste through the vessel's decoration and vied to create the most extravagant and eye-catching ships in order to attract passengers amidst fierce competition. Specialist steamboat joiners fitted the interiors under the direction of professional decorators, who may have referred to architectural pattern books, such as Asher Benjamin's *The Architect; or, Complete Builder's Guide*, published in Boston, for their overall schemes and for decorative details.

The powered fretsaw and the power loom were among the main instruments in realizing these dream palaces, which were grander than almost any of the buildings ashore. By the 1820s gingerbread woodwork could be turned out quickly by the yard in power-mills, allowing extensive repetition of the most ornate features. Similarly, the industrial power looms that came into use early in the nineteenth century could make the acres of floral carpeting required by these boats relatively cheaply. It is an indication of the importance of style and luxury that the cost of decorating and furnishing a river steamer was typically more than half the total price of the vessel.

Like Mrs Royall, Edmund Flagg, a roving reporter on the Mississippi River in 1837, also described his observations from the paddle steamer on which he travelled:

> We were soon once more in motion; the morning mists were dispersing, the sun rose up behind the forests. We passed many little villages along the banks, and it was delightful to remove from the noise, and heat, and confusion below to the lofty *hurricane deck*, and lounge away hour after hour in gazing upon the varied and beautiful scenes which presented themselves in constant succession to the eye.[7]

These comments reinforce the sense that passengers in those more leisurely times already experienced the view from these vessels in a way that we can understand as cinematic. They, the observers, commonly described the *passing scene* as if *it* were in motion, suggesting the apparent stability of these boats.

In reality, the benign appreciation of passing scenery was not all that was going on. Inside the grand public rooms and the private cabins of river steamers, a free and easy culture rapidly developed. 'Travelling is a sad demoralizer', wrote Joseph Ingraham, reporting on the character of his fellow passengers aboard a paddle steamer bound for Natchez from New Orleans in 1835.[8] Gambling was the subject of his commentary and the 'black legs', or professional gamblers, the specific targets of his critique. 'Gambling', he wrote, 'is not only permitted but encouraged on most of the boats', as it is on board the great ocean-going cruise liners of today.

Since card games were one of the most popular pastimes of the age, the seductively styled interiors were furnished to function as casinos, enabling the slickly dressed and smoothly mannered black legs to 'take in' many innocent riverboat passengers with impunity. Elaborate techniques for cheating, employing peep-holes in cabin walls, hidden mirrors and networks of accomplices, enabled ruthless gamblers to fleece the many suckers who allowed themselves to be enticed by the heady atmosphere of freedom on the boats. The black legs were the real-life prototypes for the character Rhett Butler, the world-weary gambler-hero of the novel and film *Gone With the Wind*, which enjoyed worldwide popularity 100 years after Ingraham's account.

After the disaster of the Civil War and the consequent economic downturn for the South, the gambling culture of the riverboats became the target of legislation and was eventually banned in all the states through which the Mississippi flowed. Since 1989, however, several states have legalized gaming parlours and casinos on historic and modern riverboats, equipped with slot machines and gambling tables for black-jack, craps, roulette and poker. Following the precedent of the early riverboats and the more recent example of the great Las Vegas and Atlantic City casinos, modern riverboats present live jazz and spectacular cabaret entertainment, enhancing the festive atmosphere of a river cruise. Annual profits for the riverboat casinos have run into the hundreds of millions of dollars, making this a big regional business.

Successive generations of larger, faster and ever more elaborate steamboats, plying the Mississippi, Ohio and Hudson rivers, were fitted with interiors that corresponded to the architectural fashions of the time, many mid-nineteenth-century boats being decorated in Renaissance or Gothic Revival styles. The latter was not, however, the chaste domestic Gothic promoted at the time by the influential architect Andrew Jackson Downing, but an altogether more glamorous use of the vocabulary of Gothic architecture, repeated relentlessly through machined scrollwork and cast plaster, in public saloons up to 300 feet long. Whether inspired by the Renaissance or the Middle Ages, these vast spaces were typically painted white and gold or in vivid colours, illuminated by coloured glass windows and elaborate chandeliers, and decorated with murals of American landscapes, historical narratives or mythological subjects, the total effect generating an enormous nineteenth-century 'Wow!' factor.

The river steamboats were the great peoples' palaces of their time and offered their passengers and those who watched their majestic progress from the shore, or bought the Currier & Ives prints celebrating their competitive exploits, a spectacle unrivalled in all but the most ornate mansions of the time or in other forms of transport, such as the Pullman Palace cars. Mark Twain wrote of the Mississippi steamboats: 'to the entire populations spread over both banks between Baton Rouge and St Louis, they were palaces; they tallied with the citizen's dream of what magnificence was, and satisfied it'.[9] But in fact they were also dirty and damaged by the many rougher, buckskin-clad passengers who were inclined to whittle the furniture and bed down with their boots on.

Aboard nineteenth-century river steamers, travellers of all social classes could enjoy the theatrical backdrop of the ship's design, the gaiety of the orchestra, the free-flowing liquor and the simple intoxication of travel. Particularly to European observers, the equality among passengers of vastly different social backgrounds was notable. No one knew anyone else; and the effects of being displaced from family and home, church and community, combined with the ostentation of the ship, encouraged relaxed or even reckless behaviour among the passengers. This has remained a characteristic of shipboard life ever since.

## Transatlantic Style

In the decades that followed Dickens's voyage on *Britannia*, the expansion of inter-continental mail services and the growth of international business demanded faster ships, while steadily increasing immigration to the United States broadened the demographic of potential travellers and made bigger ships a necessity to carry large numbers in steerage class. The custom that had begun as a small part of the shipping company's income, passengers, eventually became its major source of revenue. Shipping lines competed also for the patronage of the wealthy, by offering grander and more diverting accommodations in what were still relatively crude ships.

Yet the technical development of the ocean liner advanced rapidly. Auxiliary sails and paddle wheels gradually disappeared, to be replaced by twin propellers, providing more and steadier power and increased speed, which was the greatest lure for the most demanding travellers. In the 1840s *Britannia* cruised at a maximum speed of 9 knots, whereas by 1880 the winner of the Blue Riband speed trophy for transatlantic passenger vessels was capable of 20 knots, reducing the length of a North Atlantic crossing from eighteen days to seven.[10]

By 1888 the Inman & International Line commissioned one of the first ships to present the appearance of a modern ocean liner. The *City of New York* possessed a sleek black hull, long white deckhouses and very light rigging, while the superstructure was surmounted by three tall smoke-stacks, which conveyed an image of mechanical power and speed. Passengers enjoyed electric lighting, mechanical ventilation and hot and cold running water in all cabins. The public rooms on the *City of New York* were more elaborate than any before and included a walnut-panelled first-class smoking room and a library housing 800 books. In addition to its range of accommodation in three main classes, the really wealthy traveller could hire one of fourteen private suites with bedroom and sitting room decorated in High Victorian taste, the sorts of staterooms that Dickens was led to expect on board *Britannia*. Yet even on the *City of New York* the stability and security suggested by the elaborate, hotel-like interiors were contrived to calm passengers' fears, since such ships were still highly vulnerable to the ferocity of the sea. The *City of New York*'s first-class dining saloon was illuminated in daytime by a large stained-glass dome, which was destroyed by storms more than once during the ship's service in the North Atlantic, demonstrating that an ocean liner did not have to sink in order to be dangerous.

By the end of the nineteenth century a competition had become well established among the European nations to build the grandest and fastest ship afloat. This echoed competitions to build the tallest skyscraper or the longest bridge in the world. Like the steam train, ocean liners became great symbols of iron-and-coal technology. Not surprisingly, Britain and Germany first led this race, since they were the leading naval powers in the world. France, Italy and the United States, all concerned to display not only their nation's most advanced technology, but also their taste, artistry and craft skills, followed them, from the 1920s. These floating hotels, like their counterparts on land, employed conventionally fashionable standards of taste in an effort to woo customers.

Each generation of ocean liners produced the largest moving objects on earth, extraordinary feats of architecture and engineering, which nevertheless attempted to dazzle and seduce their customers with opulent images of security in order to divert passengers' attention from the dangers and discomforts of sea travel. In 1897 the German shipping company North German Lloyd (NGL) launched the ss *Kaiser Wilhelm der Grosse*, which immediately won the Blue Riband for Germany.

This was the biggest liner afloat, with four funnels signalling its immense power and speed. It was also a style leader, with interiors designed by a single professional, Lloyd's chief architect, Johannes Poppe. Poppe's decorative style was to set the standard for the design of subsequent German liners, which became popular for their grandiosity and for the vast scale of their public spaces. The furniture was made by the firm of A. Bembé of Mainz, which was responsible for the fittings of many other vessels of North German Lloyd and the Hamburg-Amerika Line, as well as furnishing private villas for several leading German industrialists. The ship's interiors, decorated in a heavy Baroque Revival style, were highly ornamented, every available surface writhing with *putti*, garlands, shields, shells, rocks and clouds, many of them gilded, but all defying both gravity and the suspension of disbelief.

Such grandiose settings encouraged a strong sense of occasion among the passengers, ensuring that every meal or social event was an opportunity for lavish displays of clothing, coiffeur and jewellery. This, in turn, kept the hairdressers, manicurists, barbers, tailors, florists, private maids and valets, and all other support services, busy catering to the dress and grooming of first-class passengers throughout the voyage. Known as 'Rolling Billy', the *Kaiser Wilhelm* also kept the ship's doctor busy, since it was subject to terrible instability, which resulted in *mal de mer* and many bruises among the crew and passengers.

For NGL's liner *Prinz Ludwig*, launched in 1906, Poppe's interiors reflected the full extent of his taste, since he designed staterooms and public spaces in a variety of historical styles ranging from Early Renaissance to Louis XVI. Again, A. Bembé built the furniture and fittings. By 1907, however, the NGL's Director-General, Dr Heinrich Wiegand, realized that

Poppe's taste was becoming dated and turned to a group of younger artists and architects to specify the interiors of 40 cabins on a new ship, the *Kronprinzessin Cecile*. Although Poppe was still in charge of the overall design, individual cabins were executed in the geometric Jugendstil style of the Deutscher Werkstatten and Wiener Sezession by a group of innovators including Josef Olbrich, Richard Riemerschmidt and Bruno Paul. The interiors were carried out by a number of prominent firms, all hungry for the prestige of commissions to outfit ocean liners; they included Vereinigte of Munich and the Dresdener Werkstatten, both respected for their ability to deliver large quantities of well-crafted products. The Arts and Crafts aesthetic applied in these cabins signalled the possibility of a break with historicism, while maintaining a fundamental attachment to luxury and familiarity.

Class of travel profoundly affected the experience of a passenger on any ocean passage, but particularly on westward transatlantic voyages. Between 1880 and the imposition of US immigration quotas in 1920, the largest voluntary migration in human history crossed the Atlantic from Europe to the ports and factory towns of the eastern United States. Meanwhile, others poured into the western ports from the Orient. This was the first fully mechanized migration, moved by steam and iron over rails and sea. Inevitably, the ships and trains were designed in accordance with the class divisions of the time. On the great liners of the North Atlantic, second-class passengers generally enjoyed a more modest version of the pleasures offered to first-class passengers. Their staterooms may have been smaller and their public rooms somewhat less grand, but the interiors were still designed to provide comfort and pleasure. Third class, otherwise known as 'steerage' or even 'cargo class', was another story altogether. There could be no sharper contrast between the rich and the poor than in their experience on a North Atlantic passenger ship.

On some early steamships third class was housed below the water line, usually just above the deck reserved for animals. Throughout the nineteenth century most steerage passengers were in dormitories segregated by sex and in which a berth might have to be secured and held against the advances of other passengers. By 1900, however, conditions had improved. Minimum standards for third-class passengers included basic space allocations for each person, compartments for married couples and families, access to fresh air via an open deck, medical care and a diet that may have exceeded in quality what the poorer passengers had been used to at home. Cabins were cleaned daily; toilets and communal washing facilities were increased in number and quality. Third-class dining rooms were large, spartan canteens, with enamelled walls and plain furniture.

Ironically, these, rather than the grandly decorated upper-class spaces, were the interiors most admired by modernist architects in the 1920s for their efficiency and functional appearance. Despite improving physical standards, dubious moral practices were still being designed into some boats. Ethnic segregation was provided aboard the NGL liner *George Washington*, where German third-class passengers were accommodated in a section entirely separate from southern and eastern Europeans, with the intention of ensuring Teutonic 'hygiene' and 'respectability' for the former.[11]

## Sex on the Steamboat

Despite rigid class barriers enforced by the steamship companies, resourceful passengers sometimes evaded the rules. Second- and third-class passengers found various ways of de-classing themselves to enjoy the luxury and entertainment in the upper-class areas of the ship, as fictionalized in James Cameron's film, *Titanic* (1997). Meanwhile, some first-class passengers, particularly the young, were excited by the opportunity for relatively safe 'slumming' afforded by the microcosmic society of an ocean liner. The romantic potential of this activity was dramatized many times in literature and films during the twentieth century. In the Hollywood melodrama *Now Voyager*, made in 1940, a young woman from first class is discovered making love to a crew member in the luxurious and illicit back seat of a limousine stored on the ship's freight deck under an obscuring tarpaulin. This was a scandalous form of class transgression, but a tantalizing and exciting escape from the stuffy formality and exclusivity of segregated shipboard life. It also demonstrated the romantic potential of a vehicle within a vessel.

Class transgression and the unconventional attractions of the extraordinary spaces in the mechanical underbelly of a great ship were also exploited in the genre of queer literature. In Jack Fritscher's short story *Titanic* (1999), two gentlemen slip from first class below deck to the engine room of the most famous liner of all time to dally with the stokers.

> Promptly at 11, the steward led us down five flights of back stairs to the hold. The noise of the engines, only a purr in our stateroom, drowned out the sound, way above, of the orchestra playing the 'Varsouviana' . . . The maze of catwalks was lined at both rails with sailors, coalmen, cooks, mechanics and blackamoor masseurs from the Turkish steam room. The hot red tips of the crewmen's rolled cigarettes and the gentlemen's cigars blinked with each drag in the dark like stars signalling in the night. We threaded our way through the

silent, standing men, taking our bearings.[12]

Thus, the bowels of the ship are portrayed as a vast, Piranesian cruising ground.

The story of the *Titanic* disaster is so well known that it does not require repetition here; and although *Titanic* is the most famous ship in modern history, it was virtually identical to its two sister ships, *Olympic* and *Britannic*. However, the hubris of the arrogant machine's disastrous end created a legend unique in maritime history. The Olympic Class vessels, launched between 1911 and 1914, represented the pinnacle of Edwardian style and design at sea. And their aesthetic continued to dominate ocean liner design for another fifteen to twenty years, as the First World War retarded the evolution of fashionable taste and interrupted ocean liner design and construction. The interiors of the three ships were the product of a generic design strategy created for the Olympic Class vessels built by the Harland & Wolff shipyard in Belfast. Most of the surviving photographs of these interiors were taken on board the *Olympic* by the Ulster photographer R. J. Welch.[13] Public spaces and suites for first-class passengers were palatial, offering an experience akin to an abbreviated tour of grand European and Near Eastern historical architecture.

The British White Star Line's three Olympic Class vessels were the last word in size, speed and style when they were launched between 1911 and 1914. Likened to an abbreviated architectural tour of great European palaces, their interiors provided the template for luxury liners in the years leading up to the First World War and persisting throughout the 1920s.

Each space was decorated in a different period style, including Italian Renaissance, Louis xv, Louis xvi, French Empire, Adam, Georgian, Regency, Modern Dutch, Old Dutch, Jacobean, Queen Anne and Moorish.[14] Not only were the decorative schemes lavish, but also the craftsmanship and materials used were of the highest standard. Amenities included a swimming pool, gymnasium, Turkish bath, library, reading and writing rooms, various lounges and bars, veranda and palm court, a choice of dining rooms and a 500-foot-long promenade deck, exclusively for the use of first-class passengers. The opulence of these public rooms extended to the fine detailing of such technical features as the set of bronze revolving doors that prevented draughts between the promenade deck and the palm court.

According to Harland & Wolff's historian, Thomas McCluskie, the decoration of certain spaces was gendered, as would have been the case in a great house of the same period. Describing the first-class reading room, he wrote:

The pure white walls . . . and the light and elegant furniture showed that this area was specially designed to be a ladies room. Through the great bow window, which filled almost one complete side of the room, first class passengers could look out onto the Promenade deck, watch fellow passengers taking the air, and view a vast expanse of sea and sky. At the far end of the room, a large fireplace with a living fire burning cheerfully made an ideal feature of the room.[15]

Thirty-nine first-class suites, each stylistically unique, included a large private sitting room and an adjacent bedroom with a sumptuously fitted en suite bathroom. For larger parties, the staterooms could be joined together, via interconnecting doors, to form big apartments.

Second-class passengers were accommodated in two- or four-berth cabins, all with natural light, enamelled walls, linoleum floors and mahogany furniture with plush but hard-wearing moquette upholstery. Olympic Class vessels housed their steerage passengers in two-, four- or six-berth cabins finished in enamel, linoleum and painted metal. The space restrictions here precluded movable furniture. Everything was built in for maximum economy of space. Yet the third-class smoking room and lounge, with direct access to a smaller promenade deck, were comfortable, well-lit spaces, panelled in oak and furnished with simple tables, chairs and two long, elegant benches in teak. The conviviality of this facility was intended to make a potentially daunting voyage, for most immigrants in third class, a more relaxing experience.

It was these third-class sections, the open decks and the superstructures of *Titanic's* generation of liners that Le Corbusier admired, in his treatise of 1923, *Vers une architecture*, as models for a modern architectural aesthetic. This effectively reversed the conventional influence of architecture on the design of ships. He described the sun decks and

Typical elements of the ocean liner's superstructure – flat white surfaces, funnels, linear apertures, slender tubular handrails, giant cylindrical smokestacks and monumental ventilators – were absorbed into the vocabulary of modernist buildings designed by Le Corbusier and many of his contemporaries, for example (right) at the Villa Savoye, 1930.

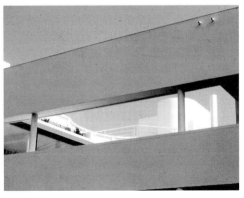

promenades of the Cunarder *Aquitania* (launched 1914) as sharing 'the same aesthetic as that of a briar pipe, an office desk or a limousine'. The rectilinear white superstructures, cabin houses and bridges of these ship provided him inspiration for the forms of his iconic white villas of the later 1920s, such as the Villa Savoye at Poissy. Picturing the *Aquitania*'s promenade lounge, simply furnished with Lloyd Loom chairs and tables, and with a ribbon of unglazed 'windows' overlooking the sea, Le Corbusier wrote:

> For architects: a wall all windows, a saloon full of light. What a contrast with the windows in our houses making holes in the walls and forming a patch of shade on either side. The result is a dismal room, and the light seems so hard and unsympathetic that curtains are indispensable in order to soften it.

He went on to suggest:

> Architects note: the value of a 'long gallery' or promenade – satisfying and interesting volume; unity in materials; a fine grouping of the constructional elements, sanely exhibited and rationally assembled . . . contrast this with our carpets, cushions, canopies, wall-papers, carved and gilt furniture, faded or 'arty' colours; the dismalness of our western bazaar.[16]

In his book *The Decorative Arts of Today*, he illustrated a ship's cabin, but it was not one of the de-luxe suites in first class. Instead it was a third-class cabin, shown for the simplicity and 'honesty' of its design revealed through its exposed rivets, bare beams, ventilation ducts, metal-framed wicker-seated chair and clean white walls. Most of all, the cabin is presented as a symbol of the modern world's unpretentious response to technical and social need and to the conditions of economy and hygiene. In the designs for his white Purist villas in the 1920s, Corb, like other radical modernists of his time, adapted many of the forms of ocean liner design to represent the modern spirit through architecture. And he imitated, in their interior decorations, the appearance of spare efficiency that he found so impressive in the decks and cabins of the *Aquitania*. It was no accident, then, that the first meeting of the International Congress of Modern Architecture (CIAM), organized by Le Corbusier, was held on a liner in the Mediterranean.

It was only after the Paris Exposition of Decorative and Industrial Arts in 1925 that modern styles of design were employed conspicuously in the first-class decoration of the next generation of trans-oceanic liners.

Between the two world wars, shipping lines across the world – Matson in the Pacific, Hamburg Sud to South America, P&O to the Far East, Orient to the antipodes and many others – launched new liners that reflected the modern age, not only through their improved performance, amenities and external appearances, but also through their style of interior decoration and furnishing throughout the ship. The first of these to court the North Atlantic trade was the French Line's *Ile de France*, launched in 1927, its interior design influenced directly by the exhibition of 1925. Just as the exhibition represented a concerted effort to re-establish France as the international leader of contemporary taste in fashion, furniture, interior design and the decorative arts, the *Ile de France* was a mobile advertisement for French aesthetics and a national emblem of French cultural excellence. Its interior design also replaced the architectural historicism of its Edwardian predecessors.

Regardless of the embarkation port, once on board the *Ile de France* a passenger was effectively in France, and spaces like the ship's terrace café were floating versions of the great Parisian cafés, such as the Dôme or La Coupole. Designs of the major public rooms reflected the heroic geometries of the 1925 exposition's pavilions, as well as their dramatic lighting effects, luxurious furnishings and semi-abstract patterns in all ornamental elements. Interiors were decorated by many of the greatest French designers from all disciplines, including the couturier Paul Poiret, who covered the walls of the first-class dining room in fine straw reeds, arranged in patterns of palm trees and tropical vegetation.

Though neither the fastest nor the largest ocean liner of the time, the *Ile de France* was recognized immediately as the most glamorous ship afloat and became a favourite of style-conscious celebrities and movie stars. Ernest Hemingway and Maurice Chevalier favoured the ship, and their presence on board added to its lustre. Press coverage of its starry clientele helped to spread the popularity of its Art Deco style across the Atlantic and on to Hollywood, which became the style's second home through the medium of cinema set design. The ship's own celebrity and mythology were disseminated through the popular arts of the time. Dorothy Fields and Jerome Kern invoked a sarcastically romantic image of the ship in one of their most popular songs, written in 1935 for the Astaire-Rogers musical *Swingtime*:

Among the grand spaces on board the classic twentieth-century ocean liners, the promenade deck was perhaps the most unique and influential, architecturally. The fantastic vistas afforded by these vast linear rooms gave passengers a clear visual understanding of the immense size of the vessels, a sense of connection with the sea and relief from any feelings of claustrophobia.

You're just as hard to land as the Ile de France,
I haven't got a chance,
This is a fine romance.

With the imposition of strict immigration quotas by the United States in 1920, the need evaporated for transatlantic liners carrying large numbers of steerage passengers. Instead, the new liners of the 1920s, including the *Ile de France*, were designed for a growing tourist trade. The old structure of first, second and steerage class was redefined by most shipping lines as cabin, tourist and third class, the last intended to provide simple, but comfortable accommodation for the likes of students, teachers or artists, travelling for pleasure, on a budget. Most passengers were business people or those private individuals and families affluent enough to travel abroad on vacation. This situation persisted until the advent of jet travel signalled the demise of regularly scheduled ocean shuttle services.

At the same time that Art Deco was becoming the style of choice for French ocean liner interiors, in America the new generation of industrial designers was applying the pseudo-science of streamlining to vessels and vehicles of all types. In both his writing and designs Norman Bel Geddes went far beyond Le Corbusier's homage to the Edwardian liners by proposing one of the twentieth century's most radical ship designs. His project for the streamlined *Ocean Liner Number 1* was published in 1932 in his book of futurology, *Horizons*, along with the extraordinary *Airliner Number 4*, which was simply an ocean liner fitted into the streamlined form of a flying-wing.

Unlike the airy, open decks and rectilinear forms admired by Corb, Bel Geddes's ship had a fully enclosed superstructure, which blended seamlessly with the softly contoured hull, the only projections being the two swept-back funnels and the cantilevered, wing-shaped bridge. The smooth surfaces of the fully glass-enclosed deck spaces were shaped to give the ship a low drag coefficient that was intended to achieve unprecedented speeds, which in turn would generate high-velocity air currents

Bel Geddes's Ocean Liner project provided a model for Hollywood set designers working on *The Big Broadcast of 1938*. The streamlined form and sleek interiors of ss *Gigantic* also compare closely with some of today's most advanced cruise ships.

around the ship. Thus, every feature, including the lifeboats, would be enclosed in the organic envelope. Nevertheless, over the swimming pool, sand beach and tiered sun decks at the stern, a glazed roof was designed to retract telescopically in order to give swimmers and sunbathers access to an open-air experience in good weather. This unrealized project, forgotten for many years, now seems to have predicted with startling accuracy many characteristics of the twenty-first-century cruise ship.

The interior of such a ship was the setting created for the musical variety film *The Big Broadcast of 1938*, which featured a Bel Geddes-like, state-of-the-art transatlantic liner, *Gigantic*, racing against a rival ship, *Colossal*, for the North Atlantic speed record. The decks, cabins, bars and lounges where Bob Hope, Martha Raye and W. C. Fields cavorted, sang and danced were designed and decorated by Hans Dreier and F. E. Freudeman in the modernistic Hollywood style that had been developed over the previous decade mainly by European émigré architects who had settled in Los Angeles and found work designing for the film industry. Interiors were almost entirely white (in this monochrome film), with all corners of the ship's structure and furnishings rounded off in radius curves to achieve an impression of fluid space and to identify all physical elements – walls, window and door openings, and furniture – as products of the latest scientific thinking. All these were typical characteristics of the theatrical modernism of Bel Geddes's architectural interiors.

Raymond Loewy achieved more modest but practical applications of streamlining in several ship designs during the 1930s. His first scheme, for the ferry *Princess Anne*, applied smoothly rounded forms to the exterior of the ship, while the inside was typical of Loewy's retail interiors and his other transport projects of the time, which included buses, cars, aircraft and trains, and featured flowing lines in all aspects of the scheme. Windows were either horizontal slits, with radius curves at their extremities, or they were circular. Furniture followed suit with long, sleek lines and more radius curves. Surfaces and fittings were decorated with parallel, horizontal bands of stainless steel, while finishes included light wood veneers,

Bel Geddes's Ocean Liner project provided a model for Hollywood set designers working on *The Big Broadcast of 1938*. The streamlined form and sleek interiors of ss *Gigantic* also compare closely with some of today's most advanced cruise ships.

linoleum and leather. His makeover of the liner *Lurline* for the Matson Shipping Company was intended to give passengers the impression that their tropical holiday had begun when they boarded the ship. *Lurline* was used on the California to Hawaii route, and Loewy's design provided the flavour of newness and sophistication required to set off the desired atmosphere of sun, sea and relaxation. According to Loewy, subsequent interior designs for ships of the Panama Line employed a

> contemporary décor free from Louis xvi, Old English or Baroque influence. A new naval décor – reflecting the modern world, efficient, comfortable and cheerful – replaced the often old-fashioned drab and stuffy pseudo-luxury of bygone eras.[17]

Loewy's consultancy designed everything from the fitted furniture of the staterooms and cabins to the crew's uniforms and the ship's menus; and they provided a follow-up service to ensure that their designs, in use, met the client's original criteria. Tellingly, the flagship ss *Panama* was requisitioned during the Second World War and during the Normandy invasion became a luxurious mobile command headquarters for General Dwight Eisenhower, an aficionado of modern transport design.

During the harsh Depression of the 1930s, the largest and most ambitious new ships were built to compete aggressively for the dwindling custom of European high society and the American rich on the North Atlantic route to New York. The greatest among these ships were the French Line's *Normandie* and the Cunard White Star Line's flagship *Queen Mary*, launched in 1935 and 1936 respectively. *Normandie* was the biggest and fastest ship afloat. It was also the most fashionable and offered the finest food and service available at sea. A large government subsidy to the French Line ensured that the boat was conceived to be superlative in every way and the ultimate symbol of national taste. The interiors were a showcase for contemporary French art, design and craft, employing the talents of glass designer René Lalique, lacquerer Jean Dunand, furniture maker Emile-Jacques Ruhlmann, painter Jean Dupas and many of the other elite French artists, decorators and crafts people of the time.

*Normandie* exuded modernity and power through its streamlined exterior forms, smooth, uncluttered decks, aerodynamically swept bow section, and three raked funnels that decreased in height towards the stern of the vessel, accentuating the visual impression of speed. Because of the ship's enormous size, designers were able to create the grandest public spaces on any ship of the period. The rigidly formal first-class dining room was more

The restrained interiors designed by Raymond Loewy, in a post-Second World War refit for the Matson Line's Pacific flagship SS *Lurline*, according to Loewy, demonstrated 'a new naval décor – reflecting the modern world, efficient, comfortable, and cheerful'.

than 300 feet long and 25 feet high, with gigantic hammered glass wall panels by Lalique, who also created the elaborate lighting scheme, comprising two massive chandeliers, full-height columnar wall lights and twelve monumental uplighters. All of this illuminated the interior of a symmetrical, gilded bronze and marble temple.

The ship's magnificent decor was primarily a dressing for the dramatic spaces in which passengers had the opportunity to perform, as if on stage. The double-height dining room, for example, was entered at one end via a high gallery from which cascaded a grand staircase on which passengers were required to make a long, highly public descent into the room. This was an opportunity to display one's poise, dress, jewellery and escort, but it could also be a trial for those less confident of their appearance. A similar grand stairway became the spectacular, theatrical setting on which Jane Russell and Marilyn Monroe made a show-stopping entrance into the first-class dining room of a classic, though nameless, ocean liner in Howard Hawks's 1953 film version of Anita Loos's novel of transatlantic seduction, *Gentlemen Prefer Blondes*. Boarding the ship in this comedy, the sexually worldly but culturally naive central character, Monroe, upon seeing a ship's cabin for the first time, exclaims with wondrous delight and surprise: 'It's like a room! Look! Round windows!'

Despite its elegance, *Normandie* was not a commercial success, surviving for only four years. And it was not popular with everyone. *Normandie* was, after all, a product of official government patronage, which ultimately produced what some observers found to be a heavy and pompous version of the Art Deco.[18] The same has been said of *Normandie*'s leading rival, the Cunard White Star Line's *Queen Mary*.

The ss *Normandie* in dock, 1930s.

The *Queen*, however, became the most popular ocean liner of all time, vying with *Normandie* throughout the later 1930s to attain the coveted Blue Riband. The difference may be that *Queen Mary* was, from the start, a more comfortable if equally luxurious ship, its decor more suited to the middle-class tastes of most of its passengers. The interiors of the *Queen Mary* were fashionable in a very Anglo-American way. Although the ship's decorators turned away from the practice of emulating past architectural styles, their version of modernity, and in particular of Art Deco, was generally more modest and homely than the *Normandie* style.

This is particularly apparent in the solid, clubby lounge furniture used throughout the ship and in details such as the coal-effect electric fires found in the staterooms. Yet glamour was applied with a trowel in areas such as the Observation cocktail lounge, where a semicircular Macassar ebony bar was highlighted by horizontal bands of polished steel, matching those encircling the brightly painted structural columns and the huge enamelled jardinières flanking the steps up to the elevated seating area, which gave a magnificent view over the forward deck to the sea beyond.

This *Queen* offered passengers in all three classes a high degree of comfort, tempered only by the need to reduce weight in the interest of speed. To this end, many modern synthetic materials were used in place of heavier alternatives such as marble and tile. Saltwater baths were available in all cabins because this lightened the load of fresh water carried on board; and many passengers considered a hot saltwater bath a luxury unavailable on land. Designed by Arthur Davis in collaboration with the French architect Charles Mewes and the American architect Benjamin Wistar Morris, the *Queen Mary* featured interiors, such as the tiled cabin-class swimming pool, that were handsomely proportioned and elegantly detailed.

The pool was an Art Deco tour de force modelled on a Pompeian bathhouse, its double-height space articulated by simplified classical forms. In typically 1930s style, stepped, chamfered corners relieved the blocky shape of the main supporting columns. At one end of the pool was a starkly symmetrical grand staircase, providing an opportunity for the extroverted to parade in their swimsuits, since physical culture was an essential element of shipboard life. The walls were skinned in handmade tiles, and the illuminated ceiling was covered in simulated alabaster to induce a sense of lightness in this windowless, deeply internal space.

Thin and very lightweight wood veneers were used to panel the cabin-class staterooms, while tourist-class cabins were simply painted and relied on a few elegant pieces of furniture and richly patterned fabrics to create an impression of luxury and style. Third-class cabins were well presented in spite of the need to economize on their fittings

The grand interior spaces of the French Line's flagship *Normandie* represent the height of Art Deco glamour, afloat or ashore, during the 1930s. Their monumental chic contrasted with the clubby, comfortable Deco style of the *Queen Mary*. National taste was a key to customer loyalty.

and furnishings. A typical four-berth third-class cabin had two 3-foot beds with fold-down, Pullman-style berths above them. Since these cabins did not offer en suite baths, a washbasin and mirror were provided, with a foldaway desk and a movable chair completing the furnishings. Their potentially spartan appearance was relieved by the addition of abstract modern carpets and upholstery materials. Carefully designed bedside lighting and other small amenities contributed to the cabins' comfortable appearance.

Great liners were not only highly decorated, but also aspired to become floating art galleries. In this respect, the *Queen Mary* was one of the less successful examples because a committee commissioned her art. Controversial painters, including Stanley Spencer and Duncan Grant, produced works that were rejected by the committee, while numerous illustrators and other commercial artists supplied murals and pictures that lacked the élan of the decorative artwork produced for the *Normandie*. The fictional hero of Evelyn Waugh's novel *Brideshead Revisited* (1945), the artist Charles Ryder, dismissed the decor of the *Queen Mary* as a reflection of the timidity and blandness in modern British culture:

> I passed through vast bronze gates on which paper-thin Assyrian animals cavorted; I trod carpets the colour of blotting paper; the painted panels of the walls were like blotting paper too – kindergarten work in flat drab colours – and between the walls were yards and yards of biscuit-coloured wood which no carpenter's tool had ever touched.[19]

The Scottish theatre designer Doris Zinkeisen was one of many artists who designed interiors and produced murals on site during the construction of the ship, ensuring that the public spaces were conceived in the spirit of stage sets. The decorative style throughout the 'ship of beautiful woods' was an essay in British bourgeois taste, magnified enormously and filtered through the lens of Hollywood. As a member of the design team, Benjamin Wistar Morris had the responsibility of ensuring that the *Queen Mary*'s interiors would fulfil the expectations American passengers had of high-class British style, and this referred particularly to the décor of English gentlemen's clubs and country houses as they were presented in Hollywood films – wood panelling and leather armchairs – plus whimsical humour and theatricality.[20]

Like the *Normandie*, the *Queen Mary* catered to celebrities, presidents and royalty. The Duke and Duchess of Windsor used it regularly, and had their stateroom 'customized' by replacing the standard curtains and linen with alternatives in the duchess's favourite colours. The

Hollywood star Marion Davies enjoyed the ship and was photographed calling 'the Coast' on the innovative ship-to-shore telephone that came as standard equipment in first-class staterooms. The Promenade Deck was described as

> a sheltered, glass-enclosed enclave of privilege . . . crowded with strollers who were fond of its sunny warmth and protection from wind and weather. Lined with deckchairs on its inboard side, it was the place to see and be seen, and to meet the most prestigious guests on the passenger list; it was the maritime equivalent of the Via Veneto.[21]

It was there that Frank Lloyd Wright strode about purposefully in his broad-brimmed hat and cape, where Churchill would enjoy a cigar, and Noel Coward exchanged quips with Marlene Dietrich, all photographed for publication. There too, a favourite pastime was to people-watch, read or simply lounge in the adjustable folding deckchairs, manufactured by the Vono Company, establishing a taste for these simple, foldable lounge chairs that spread to private lawns, porches and parks around the world.

In addition to the fame of its more polite social pastimes, the *Queen Mary* was well known as a passion pit. Its Veranda Grill, the classiest restaurant on board, was also a late-night meeting spot, where sexual liaisons were formed quickly and often fleetingly against a backdrop of exceptionally sensuous décor, designed in gold and silver by Zinkeisen. The intimate scale of the room, arranged around a small central dance floor, was enhanced by electrically lit etched-glass balusters that changed colour in time with the music, heated windowsills giving spectacular views over the fantail deck, and a 40-foot mural featuring a bare-breasted Nubian snake charmer, for inspiration.

Such interiors represent the moment of the ocean liner's complete transition beyond the requirements of transportation and into the realms of the exotic, the escapist and the romantic. The social rituals and romantic adventures inspired by the design of this and other grand ships of the period were both products of and influences over the morals and expectations of the modern world for more than half a century, and they left a legacy of myths and images that continue to haunt our notions of glamour, adventure and progress.

## Swan Song

In early 1940, with Europe at war, the *Queen Mary*, like most of her contemporaries, was conscripted to serve as a troop carrier for the

British military. In the course of the war the ship transported nearly a million soldiers of Allied nations all around the world; speed was its main defence against enemy attack.[22] In a coat of grey camouflage paintwork, the *Queen Mary* created an impression of raw mechanical power that had been obscured by its glamorous civilian attire of black hull, white superstructure and red stacks. Inside, the transformation from luxury hotel to wartime barracks was almost surreal. Although many interior fittings were removed for the duration, most major public rooms were hastily converted to new and strange purposes. The tourist-class playroom, complete with Heather 'Herry' Perry's tropical jungle murals, became the Royal Air Force's orderly room. The glorious cabin-class restaurant was lined with long mess-room galley tables. Because the ship was now 'dry', the tourist-class cocktail bar was dispensing only medical supplies, since the room had become one of the ship's busy pharmacies.

Meanwhile, below the surface of the ocean, German submarines were seeking just such rich targets for their torpedoes. The harsh environment of the u-boat was vividly portrayed in a German television series of 1981, *Das Boot*, set in the claustrophobic interiors of a German submarine, cruising the North Atlantic in 1942. The atmosphere inside the vessel was

The *Queen Mary*, like her contemporaries, was hastily converted to serve as a troop carrier during the Second World War. Capable of carrying 2,100 passengers in peacetime, the *Queen* transported more than 16,000 troops and crew in a single exercise during the war. Sleeping in stacked berths only 18 inches apart, both troops and crew suffered considerable discomfort, and some died of heat and suffocation.

Unlike its more spacious nuclear-powered contemporaries, the interior of the diesel-electric Soviet submarine *Scorpion*, built in 1972, was a cramped, mechanistic environment for the crew, whose job was to deliver nuclear death to Western targets in the event that the Cold War would turn hot.

shown to be gritty, sweaty, evil-smelling, frightening and physically oppressive. Since large motors, batteries and fuel tanks, in addition to the cargo of armaments, consumed much of the available space in the hull of the ship, the crew areas were nightmarishly cramped, and no effort was made to conceal the mechanistic nature of the vessel. Every surface and detail of the interior was clearly part of the machinery of the ship. Even the bunks in which the crew slept were cage-like, metal-framed structures within which the crewman's body was contained for a fixed period of time before another sailor took his place in a round-the-clock rotation. Daylight was rarely seen because the ship surfaced mainly at night, and any sense of time passing during underwater operations was purely academic. External sound was weirdly muffled by the medium of the sea, while inside the ship conversations of the crew were dampened by the background noise of the churning engines. Tension and anxiety were the main atmospheric constituents in this hostile environment.

Alternatively, the American comedy film *Operation Petticoat*, made in 1958, offered a lighter, sanitized picture of life in a Second World War US submarine, yet an image that reflected a distinctive approach to the design of warships, military aircraft and other fighting vehicles. Here, all was calm and comfortable, especially in the chief officers' cabin, which resembled a very small contemporary-style hotel room. This was the sort of room in which an immaculately groomed and uniformed captain, Cary Grant, could relax between shifts. It was also the sort of interior that the Raymond Loewy consultancy began designing for the American Navy during the Second World War

> to improve the life of the crew in extreme conditions of warfare . . .
> Our goal was to limit noise and heat . . . and save money at the same
> time. We had established new standards of comfort and work
> efficiency for US Navy warships.[23]

Unfortunately, Loewy's standards were realized only for a tiny percentage of the vehicles and vessels in which American soldiers fought during the Second World War. The rest more closely resembled the harsh German model.

Submarine environments did improve considerably in the post-war period thanks largely to the development of nuclear propulsion. The first nuclear-powered submarine, *Nautilus*, was built for the US Navy and commissioned in 1955. In the aggressive climate of the Cold War, it had spawned a fleet of 27 by 1960. Because the nuclear ship could cruise underwater almost indefinitely, or as long as supplies of food and water lasted, interior space and comfort gained a new importance. One of the original *Nautilus* crewmen described its interiors as being 'like the Queen Mary'. In fact, the ship was spacious enough to provide an internal 'grand staircase', a large dining room that doubled as a cinema, and ample sleeping accommodation for the entire crew. Its relatively spacious interiors were humanized by the decorative use of wood veneers, pastel laminates and bright upholstery textiles, greatly improving conditions for the crew. The expansion of usable interior space also made possible the accommodation of a larger, lethal cargo of weapons, which was the submarine's raison d'être.[24]

Back on board the troop carrier *Queen Mary* (aka the *Grey Ghost*), cabin-class staterooms, designed to accommodate two civilian passengers, were re-equipped to sleep 21 in stacked bunks barely 18 inches (45 centimetres) above one another. The ship's mechanical services, having been designed to cope with around 2,000 passengers, were unable to circulate the air sufficiently when as many as 16,000 people, including passengers and crew, were on board, even though many of these would be accommodated on open decks. The lower interior decks were not air-conditioned, and in these sections of the ship the discomfort from overcrowding became so acute on some voyages, particularly in tropical waters, that many passengers became ill, and some died from heat exhaustion or suffocation. Bizarrely, one of the most dangerous places on the ship was the first-class swimming pool, where bunks were stacked seven high and the crowding was at its worst.

The Cunard historian Leslie Reade wrote of the world's great ocean liners that served as troop ships during the war:

> In such over-crowding of human-beings it is possible there had been no parallel since the infamous slavers of an earlier age. The essential difference was, of course, that the slavers were packed with cruelty and without intelligence and the giant steamers with a sensitive understanding of the limits of human tolerance and a meticulous knowledge of household economy on a colossal scale afloat.[25]

A few of the cabin-class suites on board were retained in their elegant pre-war condition throughout the conflict, because they were used to accommodate high-ranking government officials, uso entertainers and others travelling with high military priority. Winston Churchill used his suite frequently, since it offered the safest way to travel on the Atlantic in the midst of u-boat operations, and it also served as his command post during preparations for the d-Day invasion.

Although the *Queen Mary* had travelled all over the world during the war, following the Allied victory in Europe the ship worked mainly on the North Atlantic, first as a hospital transport, ferrying wounded American GIS home from the front, then transporting the hundreds of thousands of victorious soldiers back across the Atlantic, and finally as a gigantic floating nursery moving the GIS' 20,000 European war brides and their babies to embark on new family lives with their de-mobbed husbands in the USA. At last in 1947, following a major refitting and modernization, the *Queen Mary* returned to commercial service, like her sister ship, *Queen Elizabeth*, the *Ile de France* and other surviving liners of the inter-wars generation, all having seen active service as troop carriers during the war.

Although the major shipping lines maintained their flagship ocean liners during the 1950s and into the 1960s, their numbers and importance dwindled as the airlines took over their custom and became the main transporters of intercontinental passengers. The ss *United States* represents the last generation of ocean liners constructed for regular shuttle services. As the American merchant flagship, the *United States* won the Blue Riband in 1954 and retained the title of fastest passenger ship in the world until long after its decommissioning. Significantly, its interiors were designed by a team of women led by Dorothy Marckwald, who had become a specialist in the interior design of passenger ships during a career that began in 1930 with the decoration of a fleet of vessels belonging to the Grace Line. For the ss *United States* she designed interiors with sleek, modern lines and light, neutral colours that served as backgrounds for brightly coloured furniture and fittings in blues, greens and reds. She wrote bluntly: 'We try to use all clear colors because we think muddy colors make people seasick.'[26]

With memories of the Second World War fresh in mind, the designers of the *United States* planned the ship's interiors to be easily convertible to carry troops. Some jaded observers saw the *United States* as nothing more than a massive troopship disguised as a passenger liner until the next inevitable conflict erupted. Accordingly, the ship's furniture and all its fittings were constructed of aluminium for lightness and maximum fire-resistance in accordance with Method One Design standards imposed on US passenger ships. The new synthetic material Dynel provided colour-fast linen-like textiles that were also highly fireproof. Dynel was a new and extremely adaptable synthetic, popularized as imitation fur ('It's Not Fake Anything, It's Real Dynel!'), which Markwald had woven in combination with metal threads to provide 'sparkle' for upholstery fabrics and curtains. The ship's contemporary interiors were a genuine departure from the elaborate decorative styles of pre-war vessels and more in keeping with the understated aesthetic of Loewy, or the newer Scandinavian ships, than with the *Queens* or the *Normandie*. The unmistakable luxury of the *United States*, like an executive office furnished by Florence Knoll, unpretentious and functional, was described by the *Illustrated London News* as 'rather austere but delightful'.

Eventually only the *Queen Elizabeth 2* (QE2), launched in 1969 and a floating outpost of Swinging London decorated famously in Pop Art style, remained in regular service on the North Atlantic shuttle, although it was designed also as a cruise ship, spending its time increasingly on Caribbean and on other warm-water routes. One of the last remaining ships providing a regularly scheduled shuttle service between Europe and New York was the Soviet motor liner *Mikhail Lermontov*, launched in 1972, a ship that represented new values and a new style for serious travellers.

The *Lermontov* was a modestly sized, single-class vessel, which retained, nonetheless, the dignity and character of a classic ocean liner. Unlike the earlier three-class ships, on the *Lermontov* the fare was based only on the size and position of the cabin, and all the facilities of the ship were available to every passenger, as would become common on cruise liners. Everyone was, literally, in the same boat, and as a result this ship

The vibrant modern décor of the QE2 resulted from the successful collaboration of a group of young British artists, architects and designers who gave form to the upbeat mood of Swinging London in the 1960s.

was more of a real place, like a mixed-income community, than traditional liners, which had operated like three segregated communities of different classes.[27] The decor and furnishings were uniform throughout the ship, public rooms and cabins executed in a simple but well-finished international modern style. All of the amenities and rituals of earlier liners were present on board, but they were fully Sovietized; the cinema played only Russian films, such as Tarkovsky's *Solaris*; the library was stocked only with Russian-language books, magazines and *Pravda*; the cabaret consisted mainly of Cossack dancing; there was caviar with every meal. And so, no matter where one crossed its gangplank, the *Mikhail Lermontov* was as much Leningrad as the *Ile de France* had been Paris.[28]

## Floating Malls and Nautical Coffins

By 1970, the majority of passengers going to sea were travelling over shorter distances on ferryboats of various types. New roll-on, roll-off (RORO) ferries were a development from the railroad ferries of earlier times and represented a new response to the automobile culture of the post-war period, when increasing numbers of business and pleasure travellers wanted their car with them when they arrived at their destination. For shorter daytime voyages, such as the routes from mainland Spain to the Balearics or English Channel crossings, deck space was devoted to a cargo of automobiles and tour buses.

Shipping lines also realized that, even during a short crossing, they could supplement their earnings from tickets with retail sales; and the popularity of duty free shopping, which began in Hong Kong and spread around the world in the 1960s, fuelled this trade, influencing the design of new ship interiors, which became more shopping mall than hotel. Yet, ferries retained their distinctive character, with plenty of opportunity for passengers to roam about, enjoy the views and fresh air on their open decks, and pass the time on board pleasantly.

Such innovations, however, were not without danger. The movement of cars on and off the new ferries was made significantly easier and quicker due to the development of innovative hull shapes that allowed vehicles to enter and exit through bow and stern doors and by double-ramping systems at the ports, facilitating separate access to upper and lower vehicle decks. However, the urge to facilitate loading and unloading of vehicles led to designs that compromised the safety of the ship. This flaw was dramatized by the capsize of Townsend Thoresen's state-of-the-art RORO ferry, the *Herald of Free Enterprise*, at the port of Zeebrugge in 1987. The roll-on, roll-off ship was designed for the high-volume Dover to Calais route, and could carry 1,400 people, including

The modern, functional interior of the state-of-the-art roll-on, roll-off ferry, the *Herald of Free Enterprise*, became a death-trap for nearly 200 of the 500 passengers on board when the ship suddenly capsized in the port of Zeebrugge in 1987.

the crew. Its two car decks accommodated more than 80 automobiles, 3 tour buses and nearly 50 lorries. On the day of the accident the ship had left dock with her bow doors open, allowing the lower car deck to fill quickly with water, causing the *Herald* to capsize in just 90 seconds, coming to rest half-submerged on the port side. This was human error made possible by the absence of a designed-in warning system. According to Jeffrey Sterling, Chairman of P&O, Townsend Thoresen's parent company, 'Only a fortuitous turn to starboard in her last moments prevented her from sinking completely in deeper water.'[29] Nearly 200 passengers and crew died inside the ship from drowning, as a result of being hit by articles of furniture or cargo dislodged as the boat turned onto its side, or from hypothermia in the freezing water.

In less developed parts of the world, where ferries provide an alternative to expensive airlines or travel on poor roads, accidents have been considerably worse even than the *Herald of Free Enterprise*. In some parts of Africa, ferries are chronically overloaded and poorly maintained. In such conditions, Africa's worst maritime disaster occurred off the coast of Gambia in 2002, when the overcrowded Senegalese ferry *Le Joola*, licensed to carry 535 passengers, capsized during a storm. Of around 1,900 people crammed on board, only 60 survived, resulting in a death toll higher than that of the *Titanic*. The world's worst peacetime disaster at sea was the sinking in 1987 of the Philippine passenger ferry *Dona Paz*, which burned and sank after it collided with an oil tanker, the *Vector*, in the Strait of Tabias. Both the stricken ships and the water around them became an inferno in which most of the 4,536 victims died.

Where migrants and refugees are moving in large numbers, desperate and insistent travellers commonly cajole or bribe ticket agents, who allow them to board overcrowded ferries with household possessions well in excess of the normal baggage allowance, blocking passageways and impeding escape. Many more examples of bad practice persist around the world, giving even well-designed ferries a dubious reputation as nautical coffins.

From Scandinavia, however, well-funded and progressive shipping lines demonstrated how modern ferries could relieve the worst aspects of sea journeys and provide a safer, more pleasurable experience for the discerning passenger. The fleet of Viking car ferries designed by Knud E. Hansen for Thoresen began operating on the longer English Channel crossings in the 1960s. In the most heavily used sea-lanes in the world, they brought modern Scandinavian design to the international ferry passenger. The standard of interior design, food, service and shopping was a significant step above what had previously been offered by British ships. These were clean modern ships in contrast with the suffocating railway décor common at the time.

Hansen also designed the *Skyward* and *Starward* for Norwegian Carribbean Line in the 1960s, which were among the first purpose-designed cruise ferries. These ships prefigured the later generation of long-distance cruise liners in a number of significant ways. They were one-class vessels with a high percentage of outside cabins. They featured leisure facilities, including a trademark Skybar above the top deck, providing passengers with fantastic views from a sleek, fashionable environment. *Skyward's* top-deck bar led onto a swimming pool complex, and the ship had a casino. Most importantly, *Skyward* was built to Method One specifications and, thereby, became world class.[30]

The success of these ships encouraged the design of new and increasingly large super-ferries, which operate on longer routes around the world. They too came from the Baltic, where long routes, originating in cities such as Helsinki and Stockholm, service upwards of 11 million passengers annually. There, the Silja and Viking Lines compete fiercely to woo each other's customers on the basis of comfort and amenity, as well as ticket price.

The Silja Line's sister ships, *Serenade* and *Symphony*, built in 1990 and 1991, and each around 60,000 tonnes, are larger than all but the last and biggest of the classic ocean liners of the past. They carry more than 2,000 passengers plus cars, and offer amenities that keep pace with popular technology and fashions in contemporary leisure. They are also ecologically correct, generating low foam and waves to protect the delicate flora and fauna of the eastern Baltic. Bilge, waste water and rubbish are treated and stored aboard and then discharged ashore. Nitrogen oxide emission reduction systems are fitted, and, when in port, these ships use mains electricity, rather than running their own generators.

The interior layout of the twinned *Serenade* and *Symphony* was based around an 8-metre-wide circulation spine, a glass-roofed Promenade running 140 metres through the centre of the ship from bow to stern. All the ship's activities are located off this space and include restaurants, shops and leisure facilities, among them a large nursery with playground that enables parents to enjoy the ship and increase their freedom to shop and spend while on board. The nightclub accommodates 1,000 revellers, who dance to live music on a hydraulically operated dance floor. This is adjacent to a casino, which is one of the largest gambling facilities afloat. On the deck below are a variety of restaurants, buffets, another nightclub and a gymnasium complex with pool, health club, beauty parlour and sauna. Conference rooms, audio-visual theatres and offices serve the needs of business meetings and on-board conventions, broadening the potential uses of the ship.

Influenced by the grand-hotel atria pioneered by the American architect John Portman, in buildings such as the Bonaventura Hotel in Los Angeles, the multi-deck atria of the super-ferries were decorated for ultimate visual impact. Their gilded, glass-bubble elevators and opulent finishes create an impression of luxury and spectacle against which passengers can engage in leisurely observation, dining and shopping, quickly spending the money on which the shipping line depends for revenue. All the cabins were given large windows overlooking either the atrium or the sea. The demand for outside cabins with the best views – the Baltic shipping routes afford views of magnificent landscapes and city harbours – led to the tall, blocky design of these ships, the appearance of which stands in marked contrast to the graceful proportions of the *Queens* and other classic liners. Yet the new ships undoubtedly offer greater comfort and ease, just like a modern four-star hotel in a smart shopping mall.

Many technical improvements, most importantly stabilization, made modern ferries more comfortable and safer, especially in bad weather. However, the airlines were putting up a stiff competition with ever-lower fares. By the time the Channel Tunnel opened in 1994 to compete with the shipping lines, ferry customers were being lured on board with the promise that they would be treated to a mini-cruise while crossing the English Channel. Apart from lower fares, the pleasure of the experience was the main advantage that could be advertised, against the undoubted time saving of the tunnel connection.

The competition with air carriers also left ferry operators desperate to increase the speed of their services and to inject into the experience of travel on water a dynamic quality lacking in conventional ship types. The development of radical shapes and structures, and new methods of propulsion, followed. Between the late 1960s and 2000 Hoverspeed operated several large hovercraft, the largest capable of transporting 40 cars and more than 380 passengers on the Dover to Calais route and to the Isle of Wight. The hovercraft was invented by Christopher Cockerell in the early 1950s. These fantastic vessels were propelled over the water on a cushion of fan-forced air, providing passengers an airline-type voyage, advertised as a 'flight' by the operators. The largest hovercraft in the world was the SR.N4 Mk 3, which could cruise at 100 km/h, making it the fastest ferry in operation. For safety reasons as well as economics, passengers sat in airline-style seats. Visibility was limited by spray, creating an experience akin to flying through clouds, and vibration was persistent in any conditions other than a flat calm. Yet, the excitement of travelling on an entirely new kind of vehicle and its high speed drew a steady custom to the cross-channel Hovercraft for 40 years before they were withdrawn from service.

More recently, the development of innovative hull shapes has further altered the character of ferries and the experience of ferry travel. Although never matching the high speeds achieved by the SR.N4, the Tasmanian-built Incat wave-piercing catamarans, in commercial use from ports as widespread as Hobart, Bar Harbor, Dover and Barcelona, carry 80 cars and 450 passengers at speeds up to 42 km/h. Built of aluminium and propelled by water jets, the strikingly dynamic twin-hulled vessels accommodate their passengers in a wide, low-ceilinged, multi-purpose deck-house that brings together the functions of both moving place and transporter.

This hybrid space, with continuous rows of windows around its perimeter, houses a cocktail bar, where guests can sit atop tall stools, sip drinks prepared by a barman and socialize while enjoying a panoramic view of the sea. The bar is separated by a brass rail from a cafeteria-style dining area and a large zone of lounge seating, with comfortably upholstered chairs and small round tea tables, just big enough to support a few plates of sandwiches and drinks. The remaining floor area contains rows of airline-style, semi-reclining seats for passengers who want to relax quietly or nap. A small shop and a corridor leading to the toilets complete the passenger accommodation. There is little open space in which to walk, and there is nowhere to go. This open-plan interior combines some of the traditional amenities of ocean liners with the familiar physical limitations associated with the transporter-type seating of modern intercity trains. They are excellent on many routes, but disappoint when they replace traditional ferries associated with particular places, such as the classic ferries of the Bosphorus.

The standard, contemporary décor of the Incat ferries borrows forms from industrial design (automobile interiors, airliners, trains) and commercial architecture (shops and hotel lobbies). Their style is a blend of sleek luxury and glitzy accents, such as a mirror-tiled feature, suggesting an atrium, set into the low flat ceiling above the lounge area to relieve its visual blandness. Seating is covered in two-tone upholstery, with organically shaped, ergonomically contoured insets, as found inside many recent automobiles. The result is a showy, pleasant, interior arranged for daytime voyages.

The largest Incat ships are 98 metres long and have been adapted to various military purposes: as mobile heliports, troop carriers and reconnaissance vessels fitted with sleeping facilities, lounge and office areas in the undivided cabin space. Private individuals and companies have also purchased Incats for use as pleasure yachts, their interiors being individually designed to clients' specifications. For all these purposes, the massive, single-level passenger space inside the big catamaran was ideal for customization, much like the undivided acres of floor space in a modern office building. In 1990 an Incat ship, the Seacat *Hoverspeed Great Britain*, crossed the Atlantic from New York to the Bishop Rock lighthouse in the Scilly Isles in three days, seven hours and twenty-five minutes, capturing the Blue Riband, which had been held by the SS *United States* for 35 years. Since 1990 two further Incat ships, *Catalonia* and *Cat-Link V*, have taken the Blue Riband.

Designed with speed and elegance in mind, the Seacat's big brother, the SuperSeaCat, has a steel v-shaped monohull. Its exterior was designed by Sergio Farina of the Italian car stylists Pininfarina, famous for fashioning Ferraris and Alfa Romeos, whose sensuous contours the SuperSeaCat shares. The ship's sumptuous first-class saloons, furnished under the supervision of Vittorio Garroni Carbonara, include Pullman-style reclining seats, a bar and café with panoramic views at the stern of the vessel, an observation lounge and outside deck seating, all designed to inject some poetry back into the shipboard experience. On the practical side, facilities such as telephone and fax services are meant to appeal to business passengers. Tourist-class saloons have airline-style seating on the main deck; there is also a room for mothers and babies, a 'retail therapy centre' for drinks and perfumes and facilities for disabled passengers, satisfying the needs of most currently recognized special-interest groups.

## Cruising

The idea of going to sea entirely for the pleasure of the experience began with the acquisition and conspicuous display of private yachts, originally light boats built for racing, as early as the seventeenth century by the royalty of Europe and, later, in the nineteenth century by a class of new commercially and industrially rich Europeans and Americans. The American millionaires described by Thorstein Veblen in his anthropological satire, *The Theory of the Leisure Class*, published in 1899, had already been cruising for several decades on their highly publicized private yachts, establishing a model of what the aspiring middle classes would eventually want to do with their leisure time and disposable income. Britain, however, was the most established centre of international yachting and drew annual crowds of crowned heads and the common rich to take part in the Cowes Regatta on the Isle of Wight.

There, Queen Victoria presided over the events from her 300-foot steam-powered paddle-wheel schooner, *Victoria and Albert II*, which had two suites of staterooms for the royal family, twelve cabins for the royal household and guests, a nursery, chapel, tea pavilions and several reception rooms for meetings of state business and formal receptions. The ship was used for royal relaxation and for important affairs of state, and it was loaned to visiting dignitaries on many occasions.

After the death of the queen in 1901, Edward VII, Commodore of the Royal Yacht Squadron, held summer court at Cowes on board his fast yacht, the 122-foot *Britannia*, or his comfortable yacht, the 430-foot *Victoria and Albert III*, entertaining his royal relatives from Russia or Germany and his yachting pals from around the world. The *Victoria and Albert III* normally employed a crew of 367, which was supplemented by additional staff from the royal palaces when required. Inside and out, the royal yachts of the world's monarchies were symbols of national pride and prowess, flying prominently the insignia of their royal families and decorated in the historical styles of their country of origin.

The atmosphere of a gala shipboard event, during the annual Cowes Regatta week, is well conveyed by the painter of fashionable modern life, James Tissot, in *The Ball on Shipboard* of 1874. This imaginary scene shows the festive atmosphere against which the serious business of competitive sailing took place, where the bourgeoisie played and flirted against the magnificent backdrop of the covered deck spaces on a large vessel, decorated for the occasion with brightly coloured flags of all the competing nations. These, however, were far outshone in the painting by the elegant and elaborate costumes of the ladies on board in their summer white or pastel dresses trimmed with black and topped with jaunty straw hats or floral bonnets. A congenial and relaxed atmosphere pervades the entire scene, yet it suggests the possibility of romantic liaisons, however blasé the chic women on board appear. The finery on display speaks of the fantastic investment in clothing required of those taking part in high society during the nineteenth century and the first half of the twentieth.

Since then, seafaring apparel has remained important and distinctive, although considerably less formal and more sporting. The documentary film *Jazz on a Summer's Day* (1958), directed by the fashion photographer Bert Stern, is a latter-day cinematic cousin of the *The Ball on Shipboard*. Stern used the backdrop of the America's Cup Race of 1958 to examine the appearance, dress and manners of the boat crews, comparing them with visitors to the annual jazz festival, which coincided with the regatta held in the summer resort of Newport. The sophisticated, cosmopolitan costumes of the jazz lovers, seen listening and dancing to the coolest and hottest music of the time, contrasted with the functional, seafaring attire and the tightly choreographed labour of the crew members on the decks of the elegant 12-metre yachts competing for the coveted trophy. Yet the intercut depictions of the two simultaneous

EVERY possible device for ease of operation, convenience and apparel care—they're all in a Hartmann Trunk. But this trunk is not sold on the strength of its visible features—the things you can see. It's sold on the built-in honesty and know-how manufacturing ability that's born of two generations of sincere craftsmen. Sold only by dealers who believe in unusual quality at usual prices.

HARTMANN TRUNK COMPANY, Racine, Wisconsin

The exemplary utility, simplicity and elegance of the classic steamer trunk matched the genuine functionalism of the ocean liner. Le Corbusier promoted this connection to support his architectural philosophy. Yet the advertisement above portrays the trunk as the accessory to an elaborate, historically decorated ship interior like those denounced by Corb.

events demonstrated also the unity of modern leisure in its connections: between sport and music, athletic competition and artistic endeavour, utility and fashion.

Before the Second World War, those travelling for pleasure on either a yacht or a liner would make several complete changes of costume each day, from flannel in the morning, to silk at teatime, and satin at night. The luggage required to carry all this, in addition to corsets, bustles, enormous hats and other accessories, was an industry in itself. Manufacturers such as Asprey in England, Ernst Lange in Germany, Hartmann in the USA, Hermès and Louis Vuitton in France all specialized in the production of innovative luggage that was akin to high-quality furniture and which stood imposingly in the cabins of ocean liners and yachts around the world. Their wardrobe trunks, made of lightweight wood and metal, covered externally in fine-grained leather and lined with linen or satin, opened to reveal hanging space, drawers for lingerie, special baskets for hats, hinged bins for shoes, and much more. Such portmanteaux and other smaller pieces of specialized luggage were not only impressive status symbols, they also featured in the architectural theory espoused by Le Corbusier. He identified them as exemplars of fitness for purpose and quality without pretension, both of which attributes he associated closely with travel and the design of transport vehicles. He then translated their spatial qualities and a selective version of their appearance into designs for architecture and domestic furniture. Alongside illustrations of Hermès suitcases and sports bags and an 'Innovation' trunk, he wrote:

Type-needs, type-functions, therefore type-objects and type-furniture . . . We notice among the products of industry articles of perfect convenience and utility, that soothe our spirits with the luxury afforded by the elegance of their conception, the purity of their execution, and the efficiency of their operation. They are so well thought out that we feel them to be harmonious, and this harmony is sufficient for our gratification.[31]

Yet the interiors of private yachts, like the first-class accommodations on ocean liners, usually flew in the face of Corb's admiration. They became ever more elaborate in their owners' attempts to outshine each other, whether they were European royals, Latin American presidents or North American steel moguls. In the years before the Stock Market Crash of 1929, new millionaires, making their social mark among the cream of American high society, a group of families known as the Four Hundred, built yachts calculated to impress each other as well as the rest of the world. Among these ships, the most sumptuous included the series of four progressively larger *Corsairs* built for J. P. Morgan, Pierre Lorrilard's *Rhoda* and John Jacob Astor's *Nourmahal*.

These ships afforded some of their owners the comforts they could not derive from even the grandest of their mansions. In 1906 the publisher Joseph Pulitzer commissioned the 300-foot *Liberty*, which boasted all the typical luxuries, including a music room and a gymnasium. But it offered much more to Pulitzer, who was losing his eyesight. He wrote of the yacht:

> I love this boat. Here I am at home and comfortable. In a house I am lost in my blindness, always fearful of falling on stairs or obstacles. Here, the narrow companionways give me safe guidance and I can find my way about alone. Nothing in my life has given me so much pleasure.[32]

The largest yacht to be built in an American shipyard was the *Delphine*, commissioned by the automobile magnate Horace Dodge and launched in 1921 at a cost of more than $2 million, an obscene amount of money at the time. The 257-foot steamship was a familiar sight on Detroit's 'millionaires' row' in the suburb of Grosse Point, where a deep-water channel was dug in order to moor the ship in Dodge's backyard. The interiors, furnished by Tiffany & Company, included a master cabin 25 by 30 feet (8 by 10 metres), a saloon that featured a pipe organ, eight further suites, a smoking room, card room, sick bay and promenade deck. A crew of 54 staffed the ship as it cruised the Great Lakes, the East Coast resorts of Newport and Bar Harbor, the Caribbean Islands and Hawaii. Like many of the larger private yachts, along with the great ocean liners of the day, *Delphine* was requisitioned by the us Navy during the Second World War for use as a patrol boat. The yacht was eventually restored for private charter, carrying 26 guests and a crew of 30 in supreme luxury. She sails from southern French ports, and her image has been used to advertise Renault cars, equating the sumptuous interiors of their top-of-the-line product with the luxury of deeply cushioned wicker armchairs sitting majestically on the fantail deck of the boat.

The largest sailing yacht ever built was the four-masted, square-rigged barque *Hussar*, constructed in Germany for the American stockbroker E. F. Hutton and his wife, Marjorie Merriweather Post, heiress to the Post cereal fortune. She designed the ship's interiors in a variety of historical styles filtered through the lens of Syrie Maugham or Elsie de Wolfe, the 'white ladies' who were two of the most influential society decorators in Britain and the US during the 1920s and '30s. The ship was launched in 1931, in the depths of the Great Depression, and its grandeur stood in razor-sharp contrast to the austerity of the time.

Post devoted more than two years to the project of designing the ship's layout, decor and furnishing, selecting and arranging the highest-quality antiques for the seven stateroom suites, the saloons, bar and lounges. The Huttons used the ship for pleasure cruising and for business until their divorce in 1935, when Post married Joseph E. Davies. The ship was renamed *Sea Cloud* and became a floating embassy, moored in Leningrad, since Davies was the American Ambassador to the Soviet Union. Bizarrely, Post's elaborately decorated interiors became the elegant setting in which the Soviet elite met American diplomacy against the backdrop of Stalinist purges and high-communist intrigue.

The master bedroom suite was all white, decorated in Louis XVI style, with moulded ceiling, panelled walls, canopied bed, white Carrara-marble fireplace, and gilded, swan-shaped taps in the bathroom. Another stateroom was decorated in the American Colonial style, which had been popular in domestic architecture for decades but was given added archaeological conviction by the opening in 1924 of an American Wing by the Metropolitan Museum of Art in New York. Knotty pine panelling, eighteenth-century American antique furniture with tapestry coverings, and brass lamp stands produced a cosy atmosphere, despite the room's grand scale. The only feature that confirms this as a ship's cabin, rather than a large bedroom in a Newport 'cottage' or a Park Avenue mansion, is the porthole windows. Like *Delphine*, *Sea Cloud* is now a cruise ship of the highest specification and aimed at the top end of the luxury cruise market.

Modernism was very slow to find favour in yacht interiors. Yet the early twenty-first century's most extraordinary yachts, many of them built by Italian companies, are finally establishing a unity between their advanced construction, high-tech propulsion systems, satellite navigation aids and the furnishing of their cabins. Designed by the Italo-Australian architectural firm of Lazzarini & Pickering, the 118 WallyPower is a dramatically styled and high-powered ship with an angular form akin to a Lamborghini or Maserati supercar. Carbon fibre is used extensively in its construction, and it is propelled by water jets that can take the 118-foot ship to more than 60 km/h. The 118 WallyPower is a technical tour de force, whose advanced features are expressed through the dramatic stealth geometry of its exterior form and its ultra-high-specification minimalist interiors. If Darth Vader took up yachting, this would be his vessel.

With three en-suite staterooms and accommodation for six crew members, the 118 is a small yacht by comparison with the leviathans of the past, but its aspirations are just as high. The symmetrical deck layout features a central spine from which all interior spaces lead. A skylight runs uninterrupted the length of the ship, illuminating the below-deck accommodations. Dramatic mechanisms abound. The bow section can be lifted by three hydraulic rams to reveal a 'garage' for the tender, while topside, just forward of the glass-walled deckhouse, a sunken lounge area can be covered by a retractable sunscreen when the boat is moored or docked. Inside is a large saloon with plain timber surfaces and white upholstery, giving the space a cool simplicity. A long dining table, made of carbon fibre for lightness, can be lowered into the floor to provide an uninterrupted cabin space, demonstrating the flexibility achievable in modern yacht interiors.

The 118 WallyPower redefined the look of a grand express yacht, taking such craft into the realm of high-fashion contemporary design and away from the brass and polished timber or 'period' styles of the past. Its price in 2005 was $25 million, its striking design visually justifying the cost. In addition to specialist ship designers, noted contemporary architects and product designers, including Philippe Starck, have turned their hands to yacht interiors. Most notably, in 2002 Wally employed the British architect Norman Foster to design the sloop *Dark Shadow*, with

The striking stealth geometry of the 118 WallyPower, designed by Lazzarini & Pickering, is carried through its interiors, which combine a rigorously formal plan with highly refined minimalist furnishings. A rich palette of materials includes carbon fibre, cast aluminium and teak, creating a truly distinctive ambience.

pure white laminate surfaces and polished black floors throughout its immaculately cool interiors.

In many of the seafaring countries of the world, pleasure boating became a familiar pastime of the large middle classes during the twentieth century. The broad popularity of fishing, cruising and racing created an industry in the design and construction of all manner of boats, from small, open-cockpit sailing skiffs to cabin cruisers with galley, toilet, sundeck and sleeping accommodation for two, four, six or more people. Cabin interiors ranged from a very basic, utilitarian specification, with two bunk beds tucked into the angled space under the bow deck, to luxuriously fitted lounges with convertible furniture, sofa-beds, fold-down dining tables, and so on. Slightly larger boats accommodated a room-sized captain's cabin with shower bath and one or two further cabins for guests, who would also constitute the crew.

Some of these boats were adapted for fishing, with the addition of a flying bridge, a tall lightweight structure atop the deckhouse, from which fish could be spotted, and a low transom to facilitate the landing of big fish such as marlin or tuna. Swivelling fishing chairs, equipped with footplates to assist the fisherman, would be bolted to the stern deck facing aft. In most instances, such boats would be piloted by their owner/skipper, who would stand or sit at the wheel behind a windshield and perhaps under the cover of an arched wooden roof or a canvas 'bikini top'.

One of the best-known examples of this type of boat was Ernest Hemingway's *Pilar*, from which he fished the Gulf Stream in the inebriated company of writers, artists, movie stars, sporting personalities, his family and local friends in Key West and Havana. *Pilar*, a 38-foot Playmate cabin cruiser, built by the Wheeler Shipyard of New York in 1934, was a fast diesel-powered craft whose starkly functional appearance, with shiny black hull and brightly varnished woodwork, complemented the spartan, elegant modernism of Hemingway's literary aesthetic. 'The cabin would sleep six, with room for two more in the cockpit. The galley was bright with chrome fittings . . . she would be a great little fishing machine.'[33] The superb fighting chair in the stern, like Hemingway's harpoon gun, was by Abercrombie & Fitch. *Pilar* served as the model for the boat on which the writer's fictional artist-hero Thomas Hudson hunted German u-Boats as well as large fish in the novel *Islands In the Stream*, posthumously published in 1970. His experiences aboard the boat also inspired the novel *To Have and Have Not* (1937), a story about a bootlegger trafficking between Key West and Havana, which became a film vehicle for Humphrey Bogart in the 1940s.

In England, similar boats became the maritime equivalents of the heroic First World War Paris taxicabs and London buses, which ferried soldiers to and from the fields of battle. When, in 1940, British and Allied soldiers were trapped by German occupying forces in the French port of Dunkirk, the British enlisted all vessels over 30 feet, including private boats and their skippers, to form an armada to rescue the troops and return them across the English Channel to Kent. This became one of the most inspiring episodes of the Second World War, celebrated in films such as the propaganda weepy *Mrs Miniver* (1941) and the war epic *Dunkirk* (1958). In these films, cabin cruisers were stars of a spontaneously composed 'people's flotilla' alongside hundreds of passenger ferries, fishing boats and barges being strafed and bombed, the cabin cruisers' glamour enhanced in direct proportion to the devastation of their glossy lacquer and varnish.

National characteristics have been evident throughout the modern history of boat design. The Chris-Craft Corporation of Sarasota, Florida, was one of the best-known makers of launches, speedboats and cabin cruisers from 1874 and set a standard for quality and design among its competitors. Their sleek lines and high level of craftsmanship earned them an iconic status in twentieth-century design. Chris-Craft launches displayed a purely American notion of comfort and style, with their two-tone, leather-upholstered bench seats, white plastic steering wheels and chromium-plated hardware, their interiors resembling those in a Ford Mustang convertible.

Visually similar to the Chris-Crafts, the Italian-built Riva runabouts have for decades employed reclining seat backs and other movable furnishings to convert their sumptuous, pastel-upholstered cockpits into expansive sunbeds for a full Mediterranean leisure experience. Meanwhile, the elegant and very British Andrews Slipper Stern Launches, popular for day cruising on the River Thames, were furnished with curvaceous Lloyd Loom chairs, bolted to the floor of the open cockpit. These lightweight porch or café chairs, their bentwood frames covered with woven, paper-wrapped wire, seat their occupants in a stately, upright posture signifying dignity rather than languor, alertness rather than casual relaxation. They give the cockpit and the occupants of the boat a formal appearance quite different from the rakish sportiness of the otherwise similar Chris-Craft or the sensuality of the Riva launches, echoing the very different impressions made by an Alfa Romeo, a Rover and a Buick.

At the time of writing, the United States leads the world in pleasure-boat construction and use, followed by Italy and France. France is the largest manufacturer of sailboats worldwide, mainly due to the success of Bénéteau, the world's biggest sailboat builder.[34] Wherever they are built, the interiors of modern sailing boats generally follow similar patterns to those of powerboats, with the provision of convertible berths

(opposite, top left) The elegant and very English Andrews Slipper Stern Launches (designed 1912) are typically furnished with Lloyd Loom chairs,

The Chris-Craft Super Sport (1965) with seats, steering wheel and instruments closely resembling those of a contemporary American pony car.

(right) The seductive cockpit of an Italian Riva Ariston (1968), designed and manufactured by Carlo Riva. Riva boats typically converted into luxurious and sensuous floating sunbeds.

and tables, galley kitchens, enclosed head and shower cubicles. The layout of their deck spaces, however, is defined more rigorously by the parameters of wind propulsion. Whereas the powerboat has its motors located either outboard, on the transom, or inboard, under the floor of the deck or cabin, the masts, spars, stays, booms, winches, cleats, control lines and all other elements of rigging, which make a sailboat go, are arranged in a variety of established configurations on the deck. And this is where the work and the fun are found.

While cabin cruisers adopted a variety of styles that evolved dramatically during the past hundred years, sailboats maintained a relatively more familiar look, both inside and out. Steel and fibreglass have overtaken timber as the primary constructional materials; aluminium and nylon replaced wood and canvas for the rigging and sails of most boats; electric winches lightened the work of the crew; and satellite links have simplified navigation – yet the refined skill and coordinated effort of sailing remained at the core of the experience on both racing and cruising boats, and it showed.

The most specialized craft, such as the small, light and very fast twin-hulled Hobie catamarans, built of fibreglass and aluminium, offer sailors the thrill of proximity to the water while achieving unprecedented speeds for a small sailboat. The sling-type, canvas platform, on which the crew sits to control the boat, suspends them between the hulls and only inches above the surface of the water, providing a level of excitement unrivalled in the enclosing cockpits or the solid decks of most conventional, single-hull sailing boats. Like many other one-design racing machines, these boats compete in single-class fleets at sailing clubs around the world. Hobie Cats have helped to popularize the catamaran and trimaran boat types, exploiting the development of new materials and design innovation to heighten the experience of sailing as a sport and as a recreational pastime.

## Now Voyager

In the twentieth century, those who answered the lure of the sea, but did not have the money to own and operate a sea-going yacht, had the option of booking passage on an organized cruise. The Baptist minister Thomas Cook is often credited with founding the modern tourist industry. He and his son John arranged the first temperance tours of Britain in the 1840s, and soon they expanded their operations to include Europe. In 1872 he offered the first Round the World Tour, using steamship, train and stagecoach to transport his clients across North America, Japan, China and India. By the 1920s the major railroad and steamship lines were all offering coordinated tours of the Mediterranean, the Baltic, the romantic ports of the Caribbean, and the exotic Pacific. For the struggling shipping lines this presented new sources of income and an opportunity to employ their excess capacity in the low season or to extend the life of elderly ocean liners scheduled for retirement.

And so, people of ordinary means headed mainly for the sun, first in small numbers, which increased during the 1920s and '30s, when the glamour and ease of a cruise began to appeal to many who had the cash to spend on a pre-paid, packaged holiday at sea, with organized sightseeing visits at a variety of ports. This was a safe way to see the world on a tight budget, or even a lavish budget, but it was evident from the start that cruising would become popular among men and women looking for a shipboard romance. The promise of winter sun, moonlight shining on calm water, the opportunity to dress up for social events and games organized by a Social Director to bring single people together, all contributed to the popularity of such vacations. In Woody Allen's *Radio Days* (1987) set in late 1930s New York, a single woman of advancing age and limited funds speculates enthusiastically about her annual holiday:

What do you think? This year, should I go to the mountains or on a cruise? My dancing teacher says that there are more men in the mountains, but on a cruise they have more money.

An up-market cruise was portrayed in *Now Voyager* (1940), in which the wealthy heroine embarks on a romance during an early sail-fly holiday, a South American sea cruise from which she returns on a mighty Pan American Airways Martin M-130 flying boat. Along the way, the film exploits all the typical elements of the luxury cruise liner as background to a developing romance. The character's extensive cruise wardrobe is shown off against the sophisticated interiors and sunlit decks of the liner, while the narrative is played out before ship's railings, in deckchairs, and on a theatrical grand staircase descending into the first-class dining room. The combination of its raw mechanical aesthetics and the ultimate glamour of its interiors made the modern liner a uniquely potent setting for romance.

By the mid-1950s the cruise ship had become a familiar setting for television and film comedy. Gale Storm's American musical sitcom *Oh! Susannah* (broadcast 1956–60), the long-running television series *The Love Boat* (1977–86), and the British film comedy *Carry On Cruising* (1962) presented the ship as secure, benign, even cosy, a place in which the potential for humorous romance or romantic misadventure lurks around every ship's ladder and ventilation funnel.

In darker waters, disaster and adventure films have explored the labyrinthine interiors of cruise ships and liners as the extraordinary environments for extreme events. Both *The Poseidon Adventure* of 1972, filmed aboard the *Queen Mary*, and *Poseidon* (2006) chronicled the efforts of a group of passengers to escape from the hull of a cruise liner, capsized by a tidal wave, before it sinks. Throughout much of these films, all the internal spaces of the ship are filmed upside down, creating a surreal setting for the journey of the desperate survivors from the grand ballroom through the bowels of the ship, struggling against fire and rising water to reach the propeller shaft, through which the last survivors make their eventual escape onto the enormous upturned hull.

With the precipitous demise of many formerly great shipping lines in the 1960s – Compagnie Générale Transatlantique (The French Line), Union Castle, Svenska-Amerika Linea (Swedish-American Line) and many more – numerous good-quality and relatively young liners were surplus to requirement for scheduled passenger services. At first, owners remodelled these ships for the growing cruise trade operating in warm waters such as the Caribbean and Mediterranean.

The Italian liner *Leonardo da Vinci*, known as 'the floating Uffizi Gallery', was transferred from its Atlantic shuttle service to cruise the

Mediterranean from 1966–76, stopping at ports including Palma de Mallorca, Barcelona, Lisbon, Palermo and Casablanca. The *Leonardo* was also deployed on the South American cruise routes, and once it was sent through the Panama Canal to Hawaii. By the mid-1970s, however, it was clear that the ship was not well suited to its new purpose, both in terms of economy and amenity. Refitting would not cure the fundamental problems, such as the ship's inability to dock independently in shallow-water ports, and so the *Leonardo* and other similar ocean liners were eventually scrapped. It was aboard just such a grand, superannuated liner, the *Ile de France*, on its last cruise before retirement, that *The Last Voyage* (1960) was set.

With the increasing popularity of cruising in the 1990s, new ships were designed to cater to a much more demanding clientele in the mass tourism market. The new ships also had to comply with tougher safety regulations and with the evolving economic parameters of cruising. Huge new ships took on a form quite specific to the culture and routines of cruising. Hulls were designed with a shallow draft to enable the ships to reach the maximum number of desirable ports, and other technical innovations improved manoeuvrability, to make tugboat-free docking easier. As with the super-ferries, pioneered by the Baltic shipping lines, the form of the modern cruise ship has been defined partially by the high demand for outside cabins, which led to the boxy and towering architectural superstructures of these vessels. Beginning with Knud E. Hansen's design for the cruise liner *Skyward*, launched in 1969, outdoor activities were moved from open promenade decks to top-deck lidos, where swimming pools, bars and cafés offered spectacular panoramic views of the sea and passing tourist sites. Despite the rectilinear mid-sections of these boats, designers concocted extravagantly streamlined and organic elements to the bow and stern areas, creating some challenging aesthetic juxtapositions.

Like the larger super-ferries, many new cruise liners employ the spatial formula of a multi-deck atrium as the spine of the interior. Typically, these atria rise from three to ten decks in height and serve as the ship's main reception area and primary circulation space. The small size of cabins encourages passengers to spend more time in this arcade area spending money on food and entertainment. Décor ranges from classical British luxury to Las Vegas glamour, and in some cases combines both, as on board the Norwegian-registered *Brilliance of the Sea*.

Although this ship looks like the proverbial floating office block, its design is redeemed by the use of huge amounts of glass in all public areas and staterooms, creating a sense of closeness to the sea. Some of the larger spaces, such as the main dining room, emulate the luxury liners of earlier years. The Minstrel Dining Room directly apes the cabin-class restaurant of the original *Queen Mary*, with a height of two decks, a grand staircase descending from the upper level past a backlit waterfall to the main dining floor, and a small stage for a pianist. On the very top level of *Brilliance* is an observation lounge, used as a discothèque at night, featuring a revolving bar and decorated in a livery of glitz and glitter. Next door, the showpieces of the ship are its spa and the spectacularly exotic solarium, with an Indian-style pool and a retractable glass roof. Here, three giant elephants, 16 feet tall (5.5 metres), guard over the pool amid lush tropical planting and another internal waterfall. Completing the fantastic scheme, two kiosks containing the whirlpool and showers are presented as miniature Indian temples.[35]

The most ambitious cruise liner of all is Cunard's 147,600-tonne *Queen Mary 2*, the largest passenger ship constructed to date. Launched in 2003, QM2 can carry more than 3,000 passengers in 1,300 cabins and is arranged over seventeen decks. In the black, white and red livery of the Cunard Line, this ship looks more like a classic ocean liner than other cruise ships, yet inside the enormous scale distinguishes it from anything that has gone before. Restraint is a difficult word to apply to anything so grandiose, yet the QM2 is conservative in comparison to the cruise liners of companies such as the Carnival Line, which specialize in Las Vegas-style décor and bombastic visual effects contrived to awe and excite the passenger.

The Grand Lobby of the QM2 extends up through a modest six decks and provides a knowingly postmodern first impression of the ship's style. A red-carpeted double staircase sweeps down from a central upper-level balcony in two opposing curves to the main floor, from which simple white columns support a wedding-cake architecture of balconies and classically decorated wall surfaces, with Renaissance-style mouldings articulating the windows of upper-level cabins that overlook this grand space. Glass elevators that rise through the space contrast with the retro style of the structure.

British nostalgia is evoked in many public areas of the ship, where extensive use is made of darker wooden panelling and other visual clues to the Cunard heritage. Like the original *Queen Mary*, this is a 'ship of beautiful wood', with light beech the preferred finish. Models of famous historical Cunard liners are displayed in substantial glass cases in various locations, along with murals, paintings and photographs reminding the passenger of QM2's distinguished heritage. Traditional English culture is recalled in the Golden Lion bar, presented as an English country pub, while the Queen's Room is an elegant ballroom where white-gloved waiters serve afternoon tea. Even the ship's horns are replicas of those on the original *Queen Mary*, the resonant sound of which must have stirred the romantic spirit in even the most casual voyager, yet the steam that accompanies each blast on QM2 is added 'for effect'. E. B. White wrote of the original: 'I heard the *Queen Mary* blow one midnight, and the sound carried the whole history of departure and longing and loss.'[36] Today, for cruise passengers on board simply to enjoy an effortless holiday, these deeper emotions are less likely to be stirred than in the days when the ocean liner was the primary means of relocation, escape or true adventure.

The modern hotel-like staterooms incorporate all the usual modern hotel conveniences, such as large amounts of storage space, interactive television, email and mini-bar. They also have views of the sea from their large windows and private balconies, which are the feature of the modern cruise ship most important to the creation of a poetic and romantic experience for passengers. The designer of the Japanese cruise ship *Crystal Harmony*, the Italian naval architect Vittorio Garroni Carbonara, described the purpose of such balconies, to watch the sea from your bed, and explained the practical issues and challenges in designing the most effective balcony:

Combining traditional splendour with Las Vegas glitz, the *Queen Mary 2*, the largest ocean liner in the world, offers its well-heeled passengers an eclectic décor redolent of Cunard's illustrious past and featuring heroic art works that portray historic ships of the line.

> It's a structural matter. Most of the ships include the veranda within the structure and you face the sea through a big squared hole in the ship's side. The steel frame surrounding that hole reduces the visibility either in the horizontal or the vertical plan. In the *Crystal Harmony*, instead, the verandas are hinged out of the structure, are lighter and more transparent and allow an extraordinary visibility all around. Cabins are much brighter and, subsequently more cheerful. This kind of veranda is certainly more architectural and less naval than the recessed one, but this is a trend which will further develop in the future.[37]

Carbonara also noted the development of the cruise liner as a Disneyesque 'floating theme park', in which the passengers are fully occupied throughout their holidays, irrespective of any itinerary. In this spirit, the Disney Cruise Line offers such vacations on a fleet of ships designed, uniquely, for family vacations with an emphasis on play activities and décor derived from Disney's cartoon and story-telling heritage. Like the company's corporate and hotel architecture by Robert Stern,

Michael Graves and others, the Disney liners are designed in the service of entertainment and pleasure with a cartoon-like character.

At present, the cruise liner has become the final holiday destination, as opposed to being the means of getting there. But this trend can develop further. *The World* is a passenger ship, launched in 2002, that contains only luxuriously serviced apartments owned by the passengers. The ship and its residents travel the oceans of the globe very slowly, following good weather and important entertainment or sporting events – and realizing Jules Verne's concept of the ship as a gigantic playground at sea, a theme developed in his novel *Floating City* (1871) and in his story *Pleasure Island.*[38]

## Place and No Place

Globalization is often represented as a singular, one-way process, that process by which Western culture is exported to all parts of the world in a capitalist wave of social, economic and political hegemony; and this seems to be as true at the start of the third millennium as it has been through the past 300 years of international relations. Yet there are many exceptions, worthy of note, that confirm that East and West have met and merged through a more mutual process than is often recognized.

Ships were the most physically impressive instruments in the process of cultural contact before the age of the plane, and so it is not surprising that their form has been represented in the art of both Eastern and Western cultures. While the West became fascinated with the artefacts and ways of living found in the East during the period of exploration, colonization and exchange, patrons and artists in the Orient also reworked European and American artefacts and environments to indulge their own romantic and exotic ideas of the West.

In Japan and China, which opened to trade with Europe and North America during the eighteenth and nineteenth centuries, the curiosity of the upper classes for the habits and artefacts of the West was reflected most extravagantly in the art and architecture of the courts. The Beijing Summer Palace, built as a warm weather retreat for the Qing emperor and his family, has a lake, gardens, bridges, pavilions, halls and towers of the most elaborate and elegant construction, reflecting the opulent lifestyle of the privileged few during the Qing Dynasty. Rebuilt in 1893 from an earlier structure on Kunming Lake, in the grounds of the palace, a large marble and timber dining pavilion shows the Eastern fascination for a means of travel developed in the West, and in particular the American West, during the nineteenth century.

Known as the Qingyanfang Boat, this unique, 36-metre-long structure adopts the form of a Mississippi River paddle steamer. Its two-storey wooden deckhouse, painted to look like marble, sits atop a genuine marble, barge-like foundation flanked by rather undersized stone paddle-wheels. The 'boat' offered its visitors, originally the Chinese Dowager Empress Ci Xi, who commissioned the building, and her guests a roman-tic simulation of the experience enjoyed by passengers of American riverboats, including elevated views over water and shore from the upper 'hurricane deck', as described in the nineteenth century by travellers such as Anne Royall. In effect, it is also a prototype for the 'floating island' theme-park cruise ship predicted by Vittorio Garroni Carbonara in 1996.

Inside, the Qingyanfang Boat featured brightly coloured tile floors and windows of coloured glass transmitting a vibrant, theatrical light into the spacious reception room. Large mirrors were also used to reflect the movement of the water and thereby to animate the space. The white simulated marble of the wooden superstructure and its extravagantly decorated architecture imitated the appearance of the steamers on which the boat is modelled, yet the style of the structure and its ornamentation reflect a sketchy knowledge of the original and offer a peculiarly Chinese interpretation of the American Greek Revival features of genuine river-boats. The naivety of this faux design is, however, consistent with the loose application of Gothic and Classical architectural motifs in the dec-oration of the genuine Mississippi riverboats, criticized by American architectural purists and commentators on taste, who saw their elaborate embellishments and furnishing as vulgar and debased.

The exoticism of the West. The Chinese Dowager Empress built this nineteenth-century dining pavilion, set in the grounds of the Royal Palace of Beijing, in the form of a Mississippi River steamboat. Its interior was full of illusions to enhance the sense of fantasy and diversion.

The use of the Mississippi riverboat as a theme for a pleasure pavilion in China makes a fitting contrast to the many floating pagoda restaurants in cities around the world, from Hong Kong to Kiev, Amsterdam and London. These people's palaces serve ubiquitous, but still exotic Chinese food in a temple crossed with a ferryboat. They also provide diners with views of harbour, river, shoreline or, as in London's Limehouse district, commercial docks.

Returning to practical transportation, in Bangkok and the surrounding area of southern Thailand a wide variety of motorized passenger boats form the backbone of the transport network. Versatile long tail boats are used for a wide range of purposes, including fishing, diving, transporting goods and carrying passengers. The largest are 10 metres long and are powered by car or truck engines. Some of these engines are gigantic, visually impressive v8s mounted at the stern and attached to a long propeller shaft, the 'long tail', which extends behind the transom at the rear of the boat. The extended prop-shaft pivots up to allow the flat-bottomed craft to negotiate shallow canals, to avoid weeds growing beneath the surface of the water, or to pull up at a beach.

Passengers and driver are accommodated in a low-sided, flat-bottomed timber hull in which they sit two abreast on wooden seats facing the bow, the driver steering from the rear. Some of the highest-powered long tail boats can achieve 80 km/h. The hull offers little sense of enclosure, putting the occupants in very close proximity to the engine and to the surface of the water. Passengers therefore experience an intense sensation of speed when the bow lifts and they lean forward to balance against the rapid acceleration, while feeling the spray and watching the nearby scenery skimming past in a blur. The tough and skilful pilots negotiate crowded waterways, with low bridges and other obstacles, at breakneck speeds. Docking and departure are swift, with passengers jumping on or off the craft often without it coming to a full stop. Because of their speed and frequency of service, these boats are the most widely used form of commuter transportation, and they provide an element of fun for the more adventurous commuters on board. Therefore they have become popular among thrill-seeking tourists.

Yet the deafening roar of their unsilenced engines is one of the most characteristic irritants in the quiet vacation islands of southern Thailand and a contributor to the urban din of Bangkok. Known for hundreds of years as the 'Venice of the Orient', Bangkok has in recent years had many of its canals paved over for roads, which now host the highest concentration of clogged vehicular traffic in the region. A new underground rail system for the city, inaugurated in 2004, is intended to relieve the road traffic problems of the city, yet it will also compete against the once supreme long tail boats and other types of water buses, which were for decades the fastest way to cross the crowded city.

The long tail boat combines the elegance and practicality of a vernacular boat type with immense mechanical power. The long, swept-up bow protects passengers from water spray, as if it had been designed for that purpose, when in fact their diesel engines have been united with a traditional hull shape to create a hybrid means of water transport, blending the characteristics of Eastern design and timber construction with Western power and pollution. For their users, they combine practical virtues with an experience of speed and an immersion in the physical membrane of the city that is unique to the culture and environment in which they were developed.

Around the world, similar fusions of Eastern and Western technology have led to the creation of boats that reflect globalized notions of comfort in travel. The oldest type of boat still in use after 4,000 years is the dugout canoe. Dugout canoes are used around the world on all sorts of waterways and for every purpose. They range in size from tiny 3-metre-long, single-person craft to enormous 30-metre ocean-going examples paddled by 80 men. The length of these boats is strictly related to the size of trees available to the builders in their locale; and these are indigenous products created by local hand-craftsmanship transmitted from generation to generation.

The production of a dugout is a group endeavour, beginning with the felling of a suitable tree such as the Sausage and Jackalberry trees of Botswana. The tree is stripped of its bark and hollowed out by burning. The hull is then scraped and carved to its optimum thickness for strength and lightness. To achieve a form that will be manoeuvrable and comfortable, the hull is sometimes filled with boiling water and stretched laterally by the insertion of stout planks, which hold the shape until the wood is dry and stable. With the addition of sails and outriggers, the basic boat type has been adapted to use in various climates and sea conditions, and for many different purposes.

In recent times, the adaptation of this traditional craft has taken a turn that reveals some of the principal themes of late twentieth- and early twenty-first-century design thinking. The venerable British furniture designer Robin Day uncovered a particularly vivid example of this turn. Day was brought up in the English furniture-making town of High Wycombe, where he studied at the local school of art and technology, before moving on to the Royal College of Art in London. In the late 1940s he and his wife, the textile designer Lucienne Day, opened a design practice, where they applied the most creative of modern design thinking in the United Kingdom. Day won the Museum of Modern Art's 'International

The ubiquitous Polyprop chair, designed by Robin Day and produced by the millions since 1962, found an unusual purpose furnishing traditional dugout canoes of Botswana. This sort of ad hocism became a feature of transport vehicles and vessels in less developed parts of the world during the twentieth century. The Polyprop combined the virtues of cheapness, lightness, comfort and adaptability.

Competition for Low-Cost Furniture Design' and in 1949 became design director of the Hille furniture company, where he achieved his greatest success with his design for the Polyprop chair, which entered production in 1962.

The design of the Polyprop chair was influenced by the elegant and expensive shell-form chairs of Charles Eames, but Day aimed to produce an all-purpose product for extremely low-cost and high-volume mass-production. The Polyprop was the first chair with a one-piece injection-moulded seat and back, made of polypropylene, and supported on a tubular steel frame suitable for stacking. More than 14 million of them have been made since 1962, and they have been distributed all over the world. Although they were intended for the village hall and school canteen, they have had a more extensive and varied life than their designer imagined.

On a visit to Botswana, at a remote river landing, Day came upon a small fleet of traditional dugout canoes, which are usually propelled in shallow water by a standing man, using a long pole like a gondolier. In deep water, the crew members crouch in the hull of the boat while paddling. The canoes that Day found, however, were equipped with rows of Polyprop chairs fixed to the inside of the boats' wooden hulls, offering their paddlers the comfort of a flexible, waterproof seat and a new position for more efficient and less tiring work.

This elegant, ingenious and economical hybrid product represents a telling fusion of cultural values and is pregnant with contemporary design issues. These include ad hocism, the appropriation and use of a product or part of a product to a purpose its makers did not intend or foresee, and the reuse of Western products in less developed parts of the world. Reuse has been a significant issue since the early years of the twentieth century, when the world's poor or displaced began making sandals from the rubber of disused truck or car tyres. Later, it gained importance with the broad recognition of environmental issues, including global warming and land pollution. The synthesis of mass-produced elements, such as the Polyprop chair, and the indigenous craft product also illustrates one of the directions in which globalization and post-Fordist methods of manufacture may be moving. This involves the breakdown of singular or linear production methods, such as the assembly line, in favour of mixed modes of production, typified by flat-pack furniture and the IKEA model of user participation in the production process. Finally, the globalization of notions of comfort is suggested by the interesting fusion of modern, ergonomically designed seating with traditional types of work, in this case paddling a hand-hewn boat over water.

# 3 AIR

## The Aerial View

> The ground under you is at first a perfect blur, but as you rise the objects become clearer. At a height of one hundred feet you feel hardly any motion at all, except for the wind which strikes your face. If you did not take the precaution to fasten your hat before starting, you have probably lost it by this time.[1]

With characteristically cool objectivity and a trace of unexpected humour, Orville and Wilbur Wright described the sensation and imagery of early flight in a joint article, published in 1908. The aeroplane in which Orville made the first ascent in a heavier-than-air machine at Kitty Hawk, a sand-dune off the North Carolina coast, in December 1903, was a crude experiment. Yet by 1908, when the brothers published their article, they had perfected the Wright Flyer sufficiently to be granted a valuable patent and had demonstrated their machine successfully in several spectacular events, first in France and soon after in the United States. Although the delicate craft was still unstable and required the deft piloting skills of Wilbur Wright to control it safely, it had the basic attributes of all subsequent aeroplane designs.

The Flyer of 1908 was a lightweight biplane of aluminium, timber and canvas, powered by a 33 hp four-cylinder internal combustion engine linked to two propellers. It had evolved from a series of kite-like prototype gliders in which the pilot lay prone on the lower wing, as Orville had done in the first powered flight of 1903. The improved version provided side-by-side upright seating for two and was capable of attaining altitudes in excess of 500 feet in flights lasting more than two-and-a-half hours. The Wright Flyer was the wonder of its age, achieving the ancient human aspiration to fly like a bird, yet it was created by a pair of simple men, analytical in their approach to the science of invention and modest in the presentation of their accomplishment, but obsessively protective of their patents.

They demonstrated a modern character type that would henceforth be associated with aviation – the laconic pilot, 'intelligent, cold and

Herbert Bayer's visualization of airport traffic controlled by radio beams, 1943.

155

calm'.[2] Wilbur Wright was the prototype for all the later heroes of the sky – Manfred von Richthofen, Charles Lindbergh, Amelia Earhart, Chuck Yeager – and every airline pilot flying a jumbo jet today. Tom Wolfe described the professional descendants of Wilbur Wright:

> the voice of *the airline pilot* . . . coming over the intercom . . . with a particular drawl, a particular down-home calmness that is so exaggerated it begins to parody itself (nevertheless! – it's reassuring) . . . the voice that tells you, as the airliner is caught in thunderheads and goes bolting up and down a thousand feet at a single gulp, to check your seatbelts because 'it might get a little choppy'.[3]

Although Wilbur claimed to have started his aviation experiments for the fun of it, he was not a wild adventurer, but rather a practical inventor and a safe pair of hands, who could convince his passengers that it would be a bit of harmless fun to go aloft with him at the controls of his fairground contraption. Famed for his taciturn visage, he was not without imagination, humour and a poetic side to his nature, qualities, however, that were never expressed when he flew. Above all else, he was careful, allowing those who watched him or who flew at his side to indulge themselves in the fabulous aspects of flight while he concentrated on keeping the plane under control. A French journalist who flew with Wright declared: 'I have known today a magnificent intoxication. I have learned how it feels to be a bird. I have flown.'[4] Thus, while his companions succumbed to the 'intoxication of flight', Wilbur Wright remained cold sober. Since then, the sobriety and confidence of pilots and crew members on board passenger aircraft would be reinforced by the interior design of the planes' cabins to enable passengers to enjoy the pleasures of flight with a minimum of anxiety or discomfort.

From the first reliable statistics of 1908 for several years onwards, a new altitude record for controlled flight in a heavier-than-air craft was being broken nearly every month, as the technology of

Wilbur Wright and Edith Berg were photographed in the improved, twin-seat Wright Flyer of 1908. She was the first American woman to ride as a passenger in an aeroplane and may inadvertently have created the Hobble Skirt by tying a rope around her hem for modesty during the flight.

Hold on to your hat! The thin, elegant framework of the Wright Flyer provided its gentleman pilots, dressed in street clothes, a visual and sensory thrill never before known. In this illustration by Georges Scott, the Comte de Lambert views the abstract geometry of the Trocadéro as he cruises above the Eiffel Tower in 1909.

the flying machines advanced and the skill and daring of pilots increased with experience and intense competition.[5] From the seats of the Wright Flyer, which by 1908 was being built not only in Dayton, Ohio, but also in France, aviators viewed the world without the impediment of a fuselage. The pilot and passenger could look down between their feet, which were supported only by a slender tubular footrest, to see the ground pass below them, much as a modern-day hang-glider pilot sees the earth below. This was an experience of the new mechanical age that had no precedent; 80 mph wind in your face, the roar of the engine and, ultimately, a rapidly changing view of the world from ever increasing heights, 500 feet, 1,500 feet, 3,000 feet, 6,000 feet.

The illustrator Georges Scott depicted such a view, as seen by Charles, Comte de Lambert, who circled in a Wright Flyer above the Eiffel Tower and over the Trocadéro at a height of around 1,500 feet in the spring of 1909.[6] In this dramatic picture, published in the popular Parisian journal *L'Illustration*, the artist presented an aerial view of the city from the aviator's viewpoint: the majestic steel tower, the geometry of the surrounding parterre, and the visually flattened masses of housing blocks and public buildings dissected by streets and boulevards, all visible through the slender tubular structure of the aircraft. The *canard* or elevator at the front of the aeroplane spans the top of the picture, and is balanced by the pilot's trouser turn-ups and elegantly shod feet planted against the footrest at the bottom of the image. In between, the spectacle of the city is framed in quadrilateral sections by the delicate wires and thin metal tubes of the Flyer's structure.

For more than a century prior to the invention of the aeroplane, balloonists had enjoyed the thrill of ascent and spectacular aerial views. The first successful manned balloon was designed and built in France by the Montgolfier brothers, Joseph and Etienne, in 1783. Their ship carried its own fire basket above the gondola, allowing the pilots to feed the 49-foot-diameter balloon with hot air, which kept it aloft, throughout the flight. The technology of the vessel was simple, but it was aesthetically sophisticated. The huge sphere was covered in elegant Neo-classical decorations in red, gold and blue, including the royal insignia, solar images, eagles' heads and signs of the zodiac, configured like the ornamentation of a ceramic pot. The wicker gondola took the form of a narrow gallery encircling the open neck of the air bag. On its first flight, the balloon ascended from the garden of the Château de la Muette in the Bois de Boulogne to a height of 3,000 feet and travelled 7 miles before landing with a thud 25 minutes later.[7]

The following year, Joseph Montgolfier constructed a much larger balloon, La Flesselles, to carry seven people in a 22-foot-wide wicker gondola. This elaborately decorated vessel flew only once, taking its amazed and thrilled passengers to a height of 3,000 feet, from which they experienced a previously unseen vista, the raison d'être for the ascent. Writing in 1907, H. G. Wells described the luxuriously equipped 'car' of such a pleasure balloon, the basic design of which had remained effectively unchanged for more than a century:

> On the crimson padded seat of the balloon there lay a couple of rugs and a Kodak, and in opposite corners of the bottom of the car were an empty champagne bottle and a glass. 'Refreshments', said Bert meditatively, tilting the empty bottle. Then he had a brilliant idea. The two padded bed-like seats, each with blankets and mattress, he perceived, were boxes, and within he found Mr Butteridge's conception of an adequate equipment for a balloon ascent: a hamper which included a game pie, a Roman pie, a cold fowl, tomatoes, lettuce, ham sandwiches, shrimp sandwiches, a large cake, knives and forks and paper plates, self-heating tins of coffee and cocoa, bread, butter, and marmalade, several carefully packed bottles of champagne, bottles of Perrier water, and a big jar of water for washing, a portfolio, maps, and a compass, a rucksack containing a number of conveniences, including curling-tongs and hair-pins, a cap with ear-flaps, and so forth.[8]

Untethered balloons, lifted by hydrogen or hot air, offered the enjoyment of an ascent, a party and spectacular views from as high as 5,000 feet, yet their pilots had little control over the direction of the flight, apart from following prevailing air currents at various altitudes. In the nineteenth century, most of these vessels, like the Montgolfiers' ships, were brightly decorated, since they were meant to amaze and delight spectators on the ground, as well as providing spectacular sights for those on board. Their gondolas varied in size from small wicker baskets, carrying two or three persons, to large, architectural structures with space for furnishings or instruments, including large telescopes and drawing equipment.

The novelist Jules Verne portrayed such a vessel in his story *Five Weeks in a Balloon*, which followed three adventurers, financed by the Royal Geographical Society, on an exploratory flight across Africa in 1865. In this fascinating tale, he described the sensation of flight in a balloon:

> The balloon is always motionless with reference to the air that surrounds it. What moves is the mass of the atmosphere itself: for instance, one may light a taper in the car, and the flame will not even waver. An aeronaut . . . would not have suffered in the least from the speed.[9]

Verne went on to provide vivid images of the sights such a privileged group of travellers would encounter in the course of their voyages. Effects of weather featured prominently in his narrative, as did the sensations induced when the vessel rose above the clouds into thinner air:

> It was a curious spectacle – that mass of clouds piled up, at the moment, away below them! The vapors rolled over each other, and mingled together in confused masses of superb brilliance, as they reflected the rays of the sun. The *Victoria* had attained an altitude of four thousand feet, and the thermometer indicated a certain diminution of temperature. The land below could no longer be seen . . . At the height of six thousand feet, the density of the atmosphere has already greatly diminished; sound is conveyed with difficulty, and the voice is not so easily heard. The view of objects becomes confused; the gaze no longer takes in any but large, quite ill-distinguishable masses; men and animals on the surface become absolutely invisible; the roads and rivers get to look like threads, and the lakes dwindle to ponds.[10]

In their publications, the geographical societies of Europe and North America represented such elevated perspectives through illustrated articles as the technology of flight and the means of recording and publishing images advanced over the next 150 years. 'As for the appearance of Timbuctoo (sic), the reader has but to imagine a collection of billiard-balls and thimbles – such is the bird's-eye view.'[11] Eventually, such fleeting images were captured on film. In 1855 Jules Verne's friend and associate, Gaspar-Félix Tournachon, known professionally as Nadar, had patented a technique of making aerial photographs, intended for surveying and cartography. He then made the earliest aerial photograph from a balloon in 1858. The picture was a view of the village of Petit-Bicêtre, where the vessel was tethered at an altitude of 300 feet.

The eye in the sky. Nadar's balloons, the first aerial reconnaissance camera (left), today's traffic helicopters and U2 spy planes have all provided ideal positions from which to view the earth for a wide range of purposes, a service now simulated by Google Earth.

In collaboration with Verne, Nadar headed the Society for the Encouragement of Aerial Locomotion by Means of Heavier than Air Machines. Their organization was used, however, to fund the construction of one of the largest hydrogen balloons of the time, *Le Géant*. This vessel lifted a massive gondola that carried fourteen people on its first flight. The interior was equipped as an aerial photographic studio cum darkroom suitable for the wet-plate, 'collodion' photographic process that required all stages of preparation, exposure and development of the image to be executed without interruption. *Le Géant* suffered from hard landings and was not considered a success, yet Nadar continued his experiments over ten years and subsequently made further photographs from the balloon, finally perfecting a technique that eventually found new applications.

During the American Civil War, the Federal Army employed balloons with gondolas designed to observe, record and communicate troop movements, armament locations and the terrain of potential battle-fields. President Lincoln established a Balloon Corps Unit, for which Professor Thaddeus Sobieski Lowe constructed a fleet of six large hydrogen balloons, which were used in a variety of ways to obtain visual intelligence, which was telegraphed back from the balloons to military commanders on the ground.[12]

Although the Union Balloon Corps, the first 'spy in the sky', was disbanded in 1863, it had established the idea of aerial reconnaissance. Subsequent technical developments in both photography and aviation would expand the acquisition of military and civilian information from an aerial perspective, and aircraft designed as omniscient eyes eventually ranged from light helicopters, providing highway traffic reports, to the notorious, high-flying U2 spy planes of the 1960s.

Alberto Santos-Dumont, a wealthy Brazilian dandy living in Paris, advanced the capability of the balloon by creating the first controllable, non-rigid airships, eventually constructing a series of fourteen of them with which he dazzled Paris between 1898 and 1904. Santos-Dumont, bored with the limitations of uncontrolled balloon flight, devised a prototype for future airships by designing a vessel with a long, narrow, cigar-shaped gas envelope that tapered at both ends, demonstrating his awareness of rudimentary streamlining. The dirigible carried a tubular metal structure, like a bicycle frame, supporting the rudder mechanism, a passenger car and an internal combustion engine connected to a large wooden propeller. In most of his vessels the car consisted of a small wicker basket for a single person. His dirigible Number 6 carried a woven basket that tapered in from a wide, square floor to a narrow neck that

Alberto Santos-Dumont, the Brazilian pioneer aviator, designed and constructed a series of airships for his personal pleasure. This brilliant amateur was like a yachtsman of the air. His refined, aesthetic approach to flying contrasts with the Wright brothers' practical, scientific attitude towards the conquest of the air.

hugged the pilot's waist closely. Attached were the steering-control mechanism and two shallow baskets flanking the gondola to carry smaller articles. In 1901 he created a public sensation by flying from the suburb of Saint-Cloud to circle the Eiffel Tower and then returning to his starting point, a feat for which he won a prize of 125,000 francs.

Santos-Dumont pursued the art of aeronautics primarily for pleasure. His penchant for fine dining aloft, his stylish appearance at the helm and the elegance of his vessels created an elite and sybaritic image of flying that has persisted despite the development of mass travel. He commented about one of his early voyages, when he had been lunching and enjoying a good wine while cruising above the countryside near Paris: 'What dining room offered more marvellous surroundings? . . . [than] this admirable stage-setting of sun, clouds and blue sky'.[13] Santos-Dumont's fourteen dirigibles included various types, which he named according to their intended purpose and the accommodation they provided. The Racing No. 7 supported the pilot on a bicycle saddle, while, more typically, The Runabout No. 9 provided a basket for one. It was in No. 9, also known as the Baladeuse, that he allowed an American débutante, Aida de Acosta, to make the first solo flight by a woman. By contrast, The Omnibus No. 10 was designed to carry a group of partying passengers.

Santos-Dumont's dirigibles constitute a unique passage in the early development of air travel, since they were conceived almost entirely as personal transport that advanced beyond the natural limitations of the balloon, yet had not become full-blown machines, with all their attendant noise and dirt. In his dirigibles, Santos-Dumont was the skipper of an aerial motor yacht, in which he could fly home from his favourite restaurant in the Bois de Boulogne and 'moor' the ship outside his house on the Champs-Elysées. His craft could cruise at a speed fast enough to impress in the age of the horse, but slow enough to allow a leisurely perusal of the passing scenery. His altitude of around 1,000 feet was thrilling, but not terrifying. Unlike early motorists, Santos-Dumont did not contend with blow-outs, road dust or the noise of the car's unsilenced engine reverberating off the hard surface of the pavement beneath them. The innocent buzz of the dirigible's motor was dissipated in the open air surrounding the vessel and never disturbed the immaculate skipper's aesthetic contemplation.

Santos-Dumont gave up his experiments with dirigibles in 1904, when he became fascinated with heavier-than-air flight, and he experienced some success as the first aviator in France. But it was Wilbur Wright who achieved the great milestones in aviation during the first decade of the twentieth century. Not only did he and his brother fly first, but he also designed aeroplanes that could be controlled effectively, unlike those by Santos-Dumont and other pioneers. He set records for the duration of his flights, and in 1909 he took the first photograph from an aeroplane, a view over the Italian town of Centrocelli.

## Aces High

By 1909 Alberto Santos-Dumont's fame as an aviator had been eclipsed by the fabulous demonstrations of controlled flight staged in France by Wilbur Wright, and also by several other pioneers, including the brilliant American Glenn H. Curtiss, and the Frenchmen Henri Farman, Louis Blériot, Hubert Latham and Louis Paulhan, all of whom exceeded the Brazilian's modest accomplishments in heavier-than-air flight. His had been a pacific interest in the art of flying, but by the early 1910s his view was truly anachronistic, since the concept of aerial warfare had by then become deeply inculcated in modern thinking. As early as 1908, H. G. Wells had published his novel *The War in the Air*, a story that predicted the aerial bombardment of New York by German zeppelins.[14]

The world's first airline service was run before the First World War by the German zeppelin company DELAG. Their ships provided interiors as spacious and well appointed as any contemporary Pullman train, but with much more spectacular views from their large windows.

Before the technical perfection of the aeroplane during the First World War, the rigid, lighter-than-air vessels designed and constructed in Friedrichshafen, Germany, by Ferdinand von Zeppelin, were the most successful flying machines. His early airships were very large by any standards, the smallest at 420 feet in length and 38 feet in diameter, powered by two 85 hp engines and containing 367,000 cubic feet of hydrogen in sixteen gas bags for maximum lift. With an aluminium frame and silk covering, these cigar-shaped behemoths were as imposing a sight as any ocean liner of the time. In 1908 von Zeppelin had flown his LZ3 a distance of 240 miles, a trip lasting twelve hours and demonstrating the zeppelin's potential to reach England or other European destinations and to carry sufficient armaments to do real damage to any military or civilian targets.

The first zeppelins were clearly intended for military applications, and therefore the Kaiser supported their development. Yet they were also employed as civilian transport and offered luxuriously furnished passenger cabins with service to match their decor. These ships, flown by the DELAG zeppelin company, provided the world's first passenger air service. Although DELAG never actually ran scheduled flights, the company transported more than 33,000 passengers, without any civilian injuries, in 1,588 inter-city flights and tours between 1910 and 1914.[15] The three-engine ships cruised at a gentle 35–40 mph, their passenger gondolas carrying between 20 and 30 passengers in accommodations similar to the best Pullman cars of the day. Their cabins, panelled in polished mahogany plywood and equipped with lavatory and washroom, were fully carpeted and furnished with rattan and wicker chairs in groups of four, facing in pairs across a table. The tables were covered with white linen cloths on which a steward served cold meals, wine and champagne. Every seat was next to a large picture window, which opened wide for fresh air and maximum opportunity to view the sights below.[16]

In the autumn of 1914, however, the zeppelin passenger service was terminated, as the outbreak of war across Europe triggered the most dramatic leap forward in the history of aviation. Yet, in 1914, when the military applications of heavier-than-air machines were still very limited, the zeppelin had proven its flight capabilities and its potential to strike enemy targets over a very great geographical range. German military zeppelins could carry a large crew – between fifteen and thirty was common; they were equipped with a battery of powerful machine guns and hauled a heavy payload of bombs, which the ships could deliver without much interference from anti-aircraft fire.

Zeppelins provided their crews and VIP passengers with very much more austere accommodation than their luxurious DELAG predecessors. Inside the gondola, these mechanistic environments closely resembled the cabins of a warship or submarine, although much lighter, with painted metal surfaces and a bare visible structure.

In 1914 a zeppelin flying at more than 4,000 feet was safely out of the range of ground artillery. They also had the advantage of stealth, since they could cruise to their target on a prevailing wind with their engines shut off. And so, the Germans used their fleet of 117 military zeppelins for a variety of purposes, including aerial reconnaissance over the battlefields of France and Belgium. They also bombed cities, including Liège, Great Yarmouth and, most relentlessly, London, which suffered more than 200 air raids, mostly during 1915 and 1916, in which more than 500 civilians were killed and over 1,000 injured.

During the four years of the conflict, aircraft design and construction moved from its roots in independent enterprise to a vast government-funded industry backed by the huge resources of the leading combatant

By 1918 the open cockpit of a typical fighter plane had become an integrated element of the streamlined fuselage, enhancing the unity of the ace and his machine. This First World War ace is the American, Eddie Rickenbacker.

nations, Britain, Germany, Italy, France and, later, the United States. The entire nature of flying also changed, swiftly, from personal experimentation and sport to deadly aggression and military organization. The pilots who engaged in early aerial warfare developed skills and attitudes that transformed their image. A pilot came to be seen as an entirely new type of human being, one so wedded to his machine as to become part of that apparatus – a machine-man, as dreamt of by the Italian Futurist poet Filippo Tommaso Marinetti before the war and painted by Fernand Léger after it. The First World War then turned the machine-man into the 'ace'.

Between 1909 and 1914 the aeroplane had been transformed from a contraption, however elegant, of struts, wires and paper-thin surfaces, on which the pilot sat, fully exposed to the elements and as completely visible as a cavalryman on a horse, to a vehicle with an enclosing fuselage, in which the pilot was encased and partially protected. The kite-like Wright Flyer was quickly superseded by planes such as the French Morane-Saulnier Type L, designed in 1913 by Raymond Saulnier, and piloted by the first true air ace, Roland Garros. The standard, factory-issue Type L monoplane had a smooth, aerodynamic fuselage with its single radial engine placed in the nose and the 'parasol' wing suspended above the pilot, providing him with an uninterrupted view below.

Garros, however, had his plane customized in a number of significant ways, transforming it into the first effective warplane. The problem with front-engine aircraft was that pilots could not fire a machine gun straight ahead of them without their bullets destroying the wooden propeller. This single technical limitation consigned aircraft mainly to reconnaissance work in the early months of the war, leaving combative pilots to shoot at each other with hand-held carbines and pistols, while trying to control their unstable aircraft. Synchronization of the propeller and machine gun was as yet imperfect, and so Garros had his propeller armour-plated to resist damage from the stray bullets that would hit it as he fired. The observer's cockpit of the two-seat Type L was covered over, making the plane a single-seater. He also lowered and stiffened the wings for better manoeuvrability, a change that gave his Type L the appearance of a modern fighter aircraft.

With such tailor-made equipment, Garros became the most dangerous man in the sky.

Garros described using his new aircraft, spitting machine-gun fire, in an encounter with a German reconnaissance plane, armed only with a carbine rifle in the hands of the German pilot's observer. In his account, he records a phenomenon never seen before, but one that became a familiar sight in the dogfights of the two world wars and a spectacular set-piece in epic war films of the 1920s and '30s such as *Wings*, *Hell's Angels* and *Dawn Patrol*. In these films, heroic but doomed First World War pilots are seen, gloved and helmeted, trapped helplessly in the cramped cockpits of their planes as engine oil sprays them, bullets blast through the thin fabric covering of the fuselage and, finally, smoke and flame engulf them as their ships spiral to earth.

> The chase became more and more chaotic; we were now no higher than one thousand feet; at that moment an immense flame burst from the German engine and spread instantaneously. What was curious, the plane didn't fall, but descended in an immense spiral. The spectacle was frighteningly tragic, unreal. The descent became more pronounced for 25 seconds and ended with a fall of 100 feet and a horrible crash. I watched for some time to convince myself that it wasn't a dream. I carefully marked the spot and returned.[17]

While Garros had a customized plane, Allied aircraft in general were diverse, and their designs evolved quickly. German machines, on the other hand, were highly standardized at the start of the war and had the benefit of a higher specification than most Allied planes. German aircraft were equipped with a hand-operated magneto generator to start the engine, and their instrumentation included a compass and gauges for fuel, oil pressure, water temperature, altitude and speed. Some aircraft also carried a camera for reconnaissance work. Pilots had their own standard-issue pistols, hand grenades and carbines, clipped into special carriers outside the fuselage. The deep cockpits were usually fitted with padded leather or cloth upholstered seats, seat belts and triplex windshields for the comfort and protection of the pilot and an observer. Many pilots added their own personal equipment, including various sorts of lucky charms, and considered the aircraft they flew to be 'theirs'.

In a climate of rapid and relentless progress, speeds increased from the 80 or 90 mph, typical of 1914 planes, to 150 mph in 1918. Altitudes at which pilots operated also increased to more than 20,000 feet. Stunts, which had amused crowds of spectators at air meetings prior to 1914, became the techniques of dog-fighting: dives, loops, rolls and slides became the methods of gaining the upper hand or avoiding enemy fire in a dogfight, and the mastery of these skills made the aces. The increasing stability and responsiveness of newer aircraft enabled pilots to employ sophisticated tactics, such as v-formation flying, while operating their guns and maintaining panoramic observation of the traffic, both hostile and friendly, around them.

The Fokker E.1 was test flown in battle by the leading German air ace, Oswald Boelcke, who wrote of his experience in the new single-seat fighter: 'the strong man is mightiest alone. I have attained my ideal with this single-seater, now I can be pilot, observer and fighter all in one'.[18] The planes flown by Boelcke and his student cum disciple Baron Manfred von Richthofen soon became legends and were celebrated in novels and films throughout the twentieth century. Richthofen, also known as the Red Baron, developed a set of battle tactics that he put into action with his squadron of bright red Fokker triplanes, which became famous as the Flying Circus.

By the middle of the war, all combatants had the capability to chase and shoot their adversary with greatly increased accuracy. The German Fokker c Type biplane, carrying a crew of two, was the first plane to be equipped with a second machine gun, in the rear seat, enabling the crew to fire in all directions and giving them a deadly advantage over single-seat fighters. With so many 'firsts' in such a short period of time, all pilots were innovators, and a sense of adventure went along with the dangers of flying in the Great War.

The technique of bombing and the design of planes specifically for this purpose is a case in point. Early in the war, a pilot or his observer/bombardier would carry their bombs in a box and throw them by hand over the side of the plane from the safety of around 4,000 feet, an altitude out of reach of ground artillery. By 1915, with improvements in anti-aircraft guns and explosive shells, this 'safe' height had increased to more than 9,000 feet, from which only very large and prominent targets on the ground could be seen. In response, designers on both sides devised a variety of makeshift devices for carrying and dropping bombs, all of which required rapid adaptation and ingenuity from the pilot. It was not until 1916 that an effective bomb rack with a reliable mechanical release was fitted to the underside of larger existing aircraft and that specialized heavy bomber aircraft were being designed and built.

The aces operated in a new world. Their flimsy planes carried them aloft into a terrain of speed and lightness, of natural phenomena and camaraderie unlike anything known before. Physical exhilaration combined with mortal danger to produce a cocktail of previously unimaginable potency. The British pilot Cecil Lewis remembered: 'The air was our element, the sky our battlefield . . . Nobility surrounded us.'[19]

The spectatorial aspect of flying, with its own particular purposes in reconnaissance, developed alongside the art of combat and led to the design of new equipment and more specialized facilities within the aircraft. Despite the increasingly complex technology of flight, visual reference remained the pilot's primary method of navigation until the adoption of instrument reference flying around 1930. Fliers had to have excellent vision and quick responses to whatever they saw, and they had to remember what they saw. But a camera was essential for genuine aerial reconnaissance. In this respect, the greatest breakthrough of the war years was the invention, by Oskar Messter, of an automatic camera, mounted in the body of the plane, which took accurate sequential photos of the ground below, mapping large areas of terrain with no effort on the part of the pilot. Robert Wohl described the resulting image as

> a flattened and cubistic map of the earth that had little in common with the three dimensional perspective on the world that human beings were accustomed to . . . War thus brought with it a new way of seeing in which machines replaced eyes and the earth became a target as far removed from the personal experience of the observer or the bombadier [sic] as a distant planet.[20]

Aerial photos captured information such as the evidence of earlier troop movements that could not be detected by the naked eye and, therefore, belonged to a new category of image, including x-ray and microscopy photographs. Meanwhile, the art of camouflage was developed in order to conceal military installations from pilots, observers and cameras. Gun encasements were hidden under mock fairground carousels; cars and trucks were painted to blend in with the natural environment of grass and woodland or desert sand; foliage netting covered aircraft parked on the ground; and troops moved under the cover of darkness.

The ace, the machine-man, continued to evolve with the advancing technology of aircraft designed for high-altitude flying between the two world wars and into the 1940s, when the writer-pilot Antoine de Saint-Exupéry described his physical connection with his flight suit and with the mechanical services of his Lockheed p-38 Lightning:

> It takes a long time to dress for a sortie . . . There are three thicknesses of clothing to be put on, one over the other: that takes time. And this clutter of accessories that you carry about like an itinerant pedlar! All this complication of oxygen tubes, heating equipment . . .

This mask through which I breathe. I am attached to the plane by a rubber tube as indispensable as an umbilical cord. The plane is plugged into the circulation of my blood. Organs have been added to my being, and they seem to intervene between me and my heart.[21]

## 'Strap Your Belts, Please, and Close the Windows'

Following the Armistice in 1918, the future of civil aviation was seen to be in moving mail and passengers, in that order of importance. The planes used for the new civil traffic were mainly converted bombers or new planes originally designed as bombers, but adapted for civilian purposes. Since carrying mail was considered the most important aspect of the planes' work, passenger comfort was a hit or miss affair. Some of the early planes that carried delegates from London and other European cities to the Paris Peace Conference of 1919, planes such as the de Havilland DH.9, were disarmed warplanes. The DH.9 had four open cockpits, one for the pilot and three for passengers, who sat where the navigator, gunner and bombardier would have sat the previous year.

Like the pilot, passengers boarded the aircraft via a ladder. Once they took their seats, they did not move again until the end of the journey. Communication was limited mainly to sign language because of the roar of the engine. Fur-lined flight coat, leather helmet, goggles and thick gloves were appropriate apparel for London to Paris flights at any time of the year. Nevertheless, the novelty of civilian flying and the excitement of travelling in an open cockpit gave more adventuresome early passengers a memorable experience.

Lieutenant J. Parker Van Zandt, an American, reported his feelings crossing the English Channel in the open cockpit of a British-built Handley Page w8b airliner in 1925, while the other fourteen passengers sat inside the closed cabin, which contained civilities that included two rows of cushioned wicker chairs, either side of a narrow aisle, carpeted floor, electric lights and a lavatory:

I soon found myself comfortably installed in the seat of honour beside the pilot in the nose of *The Princess Mary* . . . One is scarcely conscious of the actual moment of leaving the ground until the absence of further jolting and the slight billowing effect, as the wings cushion in the warm air, make one realize with a peculiar thrill that he is off at last on his great adventure . . . Some remote atavistic instinct is stirred at this sudden break with accustomed limitations, flooding one's soul with a new spirit of power and importance, as his magic carpet bears him swiftly through the sky.[22]

The thrill of flying was not reserved for men. As a brisk young flapper in 1922, the travel writer Freya Stark crossed the Channel by airliner. She also chose a seat in the open cockpit, rather than one inside the closed passenger cabin, expressly for the thrill of flying in the open air:

> Two men lifted me . . . up over the engine . . . we flew very low. I remember a fox running below me, so it shows how very low we were. [At Le Bourget] I appealed to the nice French Customs Officer and said, 'Look, I must have a mirror before I go into Paris.' And he said 'It's only too reasonable, mademoiselle,' and a little mirror was produced and placed on the customs bench, and I was able to go tidily into Paris.[23]

The first airliners to provide all passengers a closed cabin were also planes that had been designed during the war as military aircraft, but were completed too late to take part in the fighting. Like the Handley Page w8, the Farman Goliath was a transitional type of aircraft flown by a pilot in an open cockpit sitting atop the closed passenger compartment that accommodated twelve. In early examples of this French plane, the open cockpit was not separated from the passenger cabin, which was consequently cold and windy. Furthermore, the interior space was disrupted by large diagonal cross-braces, which supported the wings and made internal circulation difficult.

Later production models had bulkheads creating two internal compartments protected from the weather. The cabins were also cleared of structural bracing and provided six rows of wicker seats fixed to the floor. The Goliath's cabin benefited from large rectangular windows, running its full length and wrapping around the nose of the plane, giving those seated in the front compartment a thrilling panoramic view. The cabins of later Goliaths, in service as late as 1933, flown by French, Dutch, Belgian, Romanian and Czech airlines, were considerably more refined, with curtained windows, tables between the seats, luggage racks and painted or inlaid wall decorations.

The architect Le Corbusier pictured the Goliath no less than eight times in *Vers une architecture* (1923), including an interior view of a Farman Air-Express version with an

The open cockpit airliner was a direct descendant of the chauffeured horse carriage and similar to the early limousine, which accommodated passengers in a closed compartment while the driver sat outside. Some of the more adventuresome early airline passengers, however, chose to sit in the open air with the pilot for maximum sensation.

Le Corbusier illustrated the Farman Goliath no less than eight times in his book *Towards a New Architecture*. The interiors of the Goliath fleet ranged from crude and spartan to elegant and refined, as shown above. The better examples were finished like high-class motorcars.

exceptionally well-appointed cabin. This example featured a flat ceiling, incorporating recessed electric lights and concealing the arched wooden structure of the fabric-covered fuselage. The furnishings are not the bare wicker chairs typical of the period, but richly tufted and buttoned upholstered seats, facing fore and aft, with matching upholstery and trim on the walls and window frames. The cabin is finished and appointed much like the interior of a contemporary Bellanger Frères limousine also pictured in the book. The Goliath was used on the prestigious London–Paris route during the mid-1920s, transporting wealthy businessmen, politicians, royals and celebrities. But its interest here remains its historic role in Le Corbusier's promotion of the International Style in modernist architecture.

During the 1920s the greatest advances in civil aviation were made in Germany, despite the severe limitations on its aircraft manufacturing imposed by post-war treaties and declarations. Nevertheless, the airline Lufthansa, which began commercial services in 1919, commissioned the design and construction of a series of monoplane airliners pioneering metal construction, advanced safety and navigational features, and passenger amenities well in advance of developments in other air-minded nations. The Fokker and Junkers companies were at the forefront of these developments with planes such as the Junkers G 24, the first three-engined monoplane of all-metal construction. From the exterior, the low wing and sleek corrugated metal fuselage of the G 24 gave it the look of a modern airliner. Passengers entered via a short set of steps through a door just behind the wing, and the plane's heated cabin was furnished with five rows of rattan- and wicker-framed, leather-upholstered chairs fixed to the floor on either side of a central aisle.

On some models, chairs with tubular steel, cast aluminium or magnesium alloy frames were used, although these were normally covered in padding of various materials with leather or cloth upholstery, concealing the mechanistic character of the frame. Some early German aircraft chairs also incorporated curved sheet-metal panels for back and seat supports, also thickly padded and upholstered for comfort. Since the Junkers factory was located in Dessau, near the Bauhaus, it is not surprising that Walter Gropius, László Moholy-Nagy, Marcel Breuer and their furniture and product design students were keenly aware of aircraft furnishings and equipment, and in some cases were involved in their design. Breuer, Mart Stam and Hans Luckhardt were among the important German modernists to design pioneering aircraft furnishings. Although most of their projects were not produced, the effect of this contact can be readily seen in their tubular steel designs for domestic furniture of the later 1920s and in the aircraft seats, designed by others, that were widely used by European airlines in the 1920s and '30s.

Every seat on the Junkers G 24 had seat belts and its own window, which could be opened. There were luggage racks, electric lighting and a lavatory at the rear. At the head of the aisle was a glazed door to the cockpit, giving passengers a view into the pilot's compartment. Lufthansa also employed stewards to look after the passengers' comfort (female flight attendants, the 'stewardess', did not arrive until the 1930s). Thus, the internal configuration and basic amenities of the modern airliner were established by 1924.

Lufthansa advanced the development of night flying, first by creating an infrastructure of guidance beacons located under the flight paths of their routes and by establishing the first training programmes for pilots in instrument-reference flying. The cockpits of their planes had illuminated instrumentation and from 1926 were fitted with radios, which could be used to guide the planes to their destination through cloud and in darkness. These were the first modern cockpits. From the passenger's perspective, such advances meant greatly improved reliability of the service, shorter travel times and considerably more convenient timetables. Radio and navigational instruments developed alongside the rapidly advancing capabilities of air traffic controllers, who quickly became essential guardians of safety.

Perhaps surprisingly, given the great distances across the North American continent, civil aviation in the United States got off to a slow start during the 1920s. This was a nation desperately in love with the automobile, and it was not until Charles Lindbergh's epochal transatlantic flight in 1927 that the country became fully air-minded, and many more people wanted to fly.[24] The plane in which most of them flew was the Ford

Tri-motor, designed by William Bushnell Stout and inaugurated in 1926.

This was a high-winged monoplane of all-metal construction similar to the Junkers G 24 and seating up to fourteen in a noisy, unheated cabin, which also suffered from severe vibration. Like most of its contemporaries, the Ford interior was similar in layout to a train compartment, but narrower and more boxlike. Large, rectangular windows ran the length of the cabin, and electric lights were provided above all seats. Passengers chewed gum to compensate for frequent changes in pressure, stuffed their ears with cotton wool to reduce the engine's roar and held on tightly to anything movable. Sick bags were provided in the event of turbulence, common in the low altitudes at which these planes flew. Henry Ford intended this plane to be an aeronautical equivalent of his mass-produced cars.

With all the potential discomfort, it may seem surprising that so many people flew at all in the early years of civil aviation, and the reasons were complex. First, facing the challenge of uncertainties and thrills while flying was attractive in itself to many adventurous individuals. One British traveller, who first flew in the late 1920s, recalled the novelty of early airline rituals:

> Courier Canfield hands us our tickets . . . and requests that we go on board . . . 'Strap your belts, please, and close the windows', he requests . . . [After take-off] The courier comes down the aisle, smiling cheerfully, and tells us we can remove the belts and open the windows.[25]

For many of those people who held an unshakeable faith in progress during the 1920s and '30s, the modernity of air travel would have been enough to outweigh minor discomforts. The privileged view of the world from the air was also a strong enticement. Gilbert Grosvenor, writing for the *National Geographic Magazine* in 1933, reported on that sense of privilege shared by those who flew for pleasure:

> There are increasing thousands who travel by air because of the entertainment and recreation obtained by visiting the marvellous new world revealed by this modern method of transportation . . . Today, comfortably seated in an armchair, anyone may soar like the eagle and within a few moments command a panorama even grander than the birds . . . The writer and his wife during the past year have enjoyed 12,000 miles of air travel in the United States and West Indies on commercial passenger planes . . . we derived a more realistic picture of the geography of the country, of the relationship

of great rivers, mountain ranges, plains, cities and islands, than years of travel otherwise afforded. From aloft one sees as one piece man-made works too large to comprehend from the ground, and one also beholds the glories of Nature, which are unsuspected by the earth-bound pedestrian.[26]

During the 1920s private aviation expanded alongside commercial flying. There was a commonly held belief that the plane would go the way of the car, and that private individuals would all, one day, have a plane as their personal transport. Then as now, it was mainly the rich who benefited from the luxury of private aviation; but sport and adventure flying, such as speed and distance competitions, took many forms and in some cases also provided a livelihood for owner-pilots. Barnstormers, who took paying passengers aloft for their first flight, operated across the US in the 1920s, flying passengers for pure pleasure in the open cockpits of First World War trainers or disarmed fighter planes. These pilots were the same thrill-seekers who performed stunts at air shows, demonstrating all the skills perfected in the dogfights of the war and some that had not previously been considered, such as wing-walking.

The sport of aviation was and is a many-faceted activity pursued for many different kinds of pleasure and purpose, but the most important during the first fifty years of powered flight were the quests for speed, altitude and distance. Glenn Curtiss, James Doolittle and Howard Hughes all held world's speed records before the Second World War. For distance and altitude competitions, some of the most successful fliers of the inter-war years flew the Lockheed Vega, designed in 1928 by Jack Northrop. Its advanced construction, a fabric-covered, moulded plywood shell with internally braced cantilevered wings, allowed for a sleek, aerodynamic form well in advance of its contemporaries. The Vega was the fastest private plane of its time, setting long-distance and altitude records, most famously with Amelia Earhart or Wiley Post at the controls.

Although it provided state-of-the-art instrumentation, the Vega's small cockpit was very simple and lacking in creature comforts. The pilot's seat was a wooden bench with separate back and seat cushions offering little comfort on long flights. The windscreen, with hand-operated wiper blades, was set high over the huge radial engine and provided limited forward visibility. Access to the cockpit was either via a small triangular door leading from the passenger cabin or up the outside of the plane and down through a panel above the cockpit. The passenger compartment, entered through a side door, seated six in close-coupled pairs, and, with no aisle, the chairs had to be folded to allow access, as in a two-door automobile.

The Sikorsky s38 Amphibion was a yacht that could fly. The pilot and passenger accommodations were separated by a glass partition that provided some acoustical privacy for the passengers without interrupting the panoramic views. Every detail of its interior was conceived to enhance the pleasure of viewing the world from the air.

Yet despite its small dimensions, the Vega's interior was elegant, its curvaceous timber structure comparable to the hull of a wooden racing yacht. From the pilot's seat, one could look inside the cantilevered wings to see their complex ribbing. Its functional beauty, speed and range made the Vega a desirable aircraft for Hollywood film stars and oil executives, as well as the transport of choice for adventurers such as Earhart, who used hers for record-breaking transatlantic and transcontinental flights, and Post, who became the first pilot to circumnavigate the globe.[27]

Another of early sport aviation's most glamorous and exotic aircraft was the Sikorsky s-38 Amphibion, a twin-engine biplane capable of operating from land or water. Designed in 1928 by the emigré Russian designer Igor Sikorsky, better known today for his invention and development of the helicopter, the s-38 was the first of a series of Sikorski flying boats commissioned by Juan Trippe for Pan American Airways and intended for commercial use in the Caribbean. Of the 111 built, many were bought privately, and among those the most memorable was owned by the naturalists Martin and Osa Johnson, writers, photographers and film-makers, who penetrated the remotest regions of the world to report on the nature, people and landscapes they found. They were, in the 1920s and '30s, what David Attenborough and Jacques Cousteau became later in the century. In 1933–4 they used their s-38, painted in zebra-striped livery, for a flying photo-safari of Africa, during which they made the first aerial photographs of Mount Kilimanjaro.

Like the Vega, the s-38 also required gymnastic skills to get in and out of, but once inside, the Johnsons enjoyed the amenities and atmosphere of a luxurious air-yacht. Continuous ribbons of windows ran along both sides of the cabin and were curtained to reduce glare. Mahogany panelling set off the rich upholstery of a three-seat sofa, which was fixed against one side wall opposite two individual wicker chairs flanking a storage cabinet, which doubled as a writing table. A built-in cocktail bar completed the furnishings. The separate pilots' compartment could be viewed through two arched windows, rather like the glass divider between the chauffeur and passengers of a formal limousine. In fact, the cockpit appears similar to the driver's seat of a contemporary luxury car, such as a Bugatti Royale or a Duesenberg. The passenger cabin also incorporated a hatch in the roof through which the Johnsons could stand to film or photograph, during flight, with an unobstructed panoramic view of the countryside below.

The s-38 was, perhaps, the most romantic sport aircraft ever produced, becoming a familiar sight in the travelogues and photographs published by the Johnsons.[28] An s-38 replica also featured in Martin Scorsese's biopic of Howard Hughes, *The Aviator* (2004), in which the hero and his movie star date take a moonlit joy ride over the glittering lights of Los Angeles, accompanied by the music of Benny Goodman and Lionel Hampton, playing *Moonglow*, and landing finally on a fairway of the Beverly Hills Country Club. Only the blasé Cole Porter could declare this a bore:

> I get no kick in a plane.
> Flying too high
> With some guy in the sky
> Is my idea of nothing to do.
> But I get a kick out of you.[29]

Flying proved to be fun for many people, but above all it offered travellers the practical advantage of speed. By the early 1930s it was clear that 75 per cent of all air travellers were business people, to whom time was money. Gilbert Grosvenor confirmed that

> commuting is already habitual with some air travellers. 'I see the same faces, over and over, week after week,' said a hostess on a Washington–Philadelphia flight . . . Of the 530,000 who bought air tickets in 1932, as against some 30,000,000 who rode the Pullmans, by far the majority *flew to save time.*[30]

Despite its crudity, the Ford Tri-motor was the first robust, reliable and commercially successful American commuter airliner, and in 1929 it instituted the first air–rail transcontinental service, from New York to Los Angeles. Although passengers took overnight sections of the journey by train in Pullman sleeper compartments, they covered most miles in the air. At first, the trip lasted 48 hours with ten stops and one overnight break in a hotel, until the introduction of night flying in 1932 cut the journey time to 24 hours. Although in the 1920s flying in general was an elite way to travel, the Fords, operated by all the major US airlines, offered a standard level of service in keeping with the modest expectations of many American business travellers at the time.

By contrast, the first truly luxurious North American service was established by Western Air Express to fly from Los Angeles to San Francisco as an experiment, funded by a Guggenheim grant in 1928, to see if passenger planes could make a profit without carrying mail, but also to convince a nervous and sceptical public that aeroplanes could feel as secure as the trains and ships to which they were accustomed. Three American-built Fokker F.10 tri-motor aircraft were employed on this route. These planes seated twelve to fourteen passengers in walnut-panelled cabins furnished with alligator-skin upholstery. Their cabins were sound-insulated, making a huge difference in the level of comfort enjoyed over the ubiquitous Ford Tri-motors. Each chair had a call button for personal service, an individual reading lamp and a folding table for work and dining. Coat cupboards, racks for light luggage and a fully equipped galley provided a level of comfort far exceeding the norm of the time and rivalling any airline in the world. These planes made a great contribution to the public perception in the US that flying could be pleasant and secure. Unfortunately, it was a Fokker F.10 in which the popular Notre Dame football coach Knute Rockne was killed in a crash following the collapse of a wing in 1931, virtually a national tragedy, which led to a ban on passenger transport in all planes with wooden wing structures.

Early European airliners had interiors designed in a variety of styles consistent with the tastes and expectations of their wealthy passengers. Travellers on European ocean liners were accustomed to those ships' highly ornate Renaissance-, Tudor- or Baroque-style cabins and public rooms, and the airlines tried to import some of

Airliner cabins in the 1920s referred to a number of existing models: the yacht or porch (Fokker F.VIIA, 1925, top), the gentlemen's club (Fokker F.III, 1921, below left) and the automobile (Junkers F13, 1923, below right), all flown by KLM Royal Dutch Airlines.

this grandeur to the tiny cabins of their aircraft. KLM's Fokker F.III aircraft of the early 1920s was flown by a pilot in an open cockpit, exposed to the elements like the chauffeur of a limousine, while the plane's five passengers sat in a closed cabin appointed like a room in a gentlemen's club, with dark wooden panelling, chintz curtains at the large rectangular windows, framed pictures on the walls, a wine table with flower vase, a fitted sofa for three people and two sumptuously upholstered movable club chairs, anchored to the floor by detachable chains. By providing such lavish interiors reminiscent of clubs, first-class hotels, Pullman carriages and ocean liners, all familiar images of reliability, security and class, the airlines hoped to calm the fears of potential flyers and demonstrate that, although they might be in an unfamiliar environment, they were at least with their own kind.

Lunch is served on a Lufthansa flight in 1933. At the same time, many long-distance planes landed for meals.

Typically, Britain persisted in the traditional decorative styling of its cabin interiors longer than any of the other leading aeronautical nations. Imperial Airways' flagship planes, the Handley Page Heracles and Hannibal class aircraft of the early 1930s, provided cabins that were the exact decorative equivalent of contemporary first-class British railway compartments, although built of lighter materials. Commodious paired seats flanking the aisle were upholstered in floral chintz; the dark wood panelled walls were ornamented with decorative inlays; and curtains were of richly patterned material, creating the overall impression of a hotel lobby or a stockbroker's living room for which no expense had been spared (see p. 13).

The 'saloons', as they were called, also offered all the amenities expected in a Pullman car at the time: generous tables between the seats were covered with linen cloths at meal times; there were overhead coat and hat-racks, two lavatories and a steward's pantry, where elaborate meals were prepared. The food and drink were also equivalent to anything found in the dining room of a top-class hotel or ocean liner. A five-course lunch or a seven-course dinner was served on Imperial's London–Paris route.

Before the widespread construction of land-based airports, flying boats provided a sensible means of expanding air routes globally. These large, luxurious ships offered a quality of accommodation excelled only by two German zeppelins, *Graf Zeppelin* and *Hindenburg*, which flew in commercial service between 1928 and 1937. Nearly matching them in glamour was the earliest prototype of the grand commercial flying boats, the

Dornier Do x, built in 1929. This twelve-engined behemoth had accommodation arranged over three decks and was capable of carrying more than 160 passengers and crew. Because of the width of its fuselage, the Art Deco interiors of the Do x, arranged as a series of rooms, were spacious and airy, the walls covered in rich floral patterns. Carpeted floors were overlaid with oriental rugs on which stylishly modern club chairs were arranged in groupings, as in a hotel lobby. The Do x made a tour around South America and visited New York to demonstrate its long-distance potential, but it was under-powered and never went into scheduled service, remaining an extraordinary folly.

Contemporary with the Do x was an unrealized project, designated 'Airliner No 4' by its creator, Norman Bel Geddes, in collaboration with the aeronautical engineer Otto Koller, and influenced heavily by the ambitious design of the Do x. Conceived in 1929 and published in Bel Geddes's futurist design manifesto, *Horizons*, in 1932, Airliner No. 4 was a flying boat proposed for the North Atlantic passenger route and intended to provide all the amenities of a great ocean liner. These included a promenade deck spanning the leading edge of the plane's enormous single wing, a games area, a dining room with stage for full orchestra, swimming pool, gymnasium, staterooms for all passengers and a dedicated cinema, on Deck Five alone – there were nine decks in all. Since the interiors of this project were drawn only in diagrammatic plans and section, one must look to Bel Geddes's subsequent work on commercial aircraft designs to imagine how the cabins of Airliner No. 4 might have appeared.

Although this project never left the drawing board, Norman Bel Geddes became one of the first consultant industrial designers to be employed by an aircraft manufacturer to design interiors for their planes. His cabins for Pan American Airways Martin M-130 flying boats, known as the Pan American Clipper and China Clippers, were executed in a practical, modern style and demonstrated his ingenious use of the complex spaces available within the highly organic fuselage of the plane. Bel Geddes used simple light-coloured plywood panelling on the bulkheads, which were arranged to create a series of large living rooms. The walls of the cabins swept up from the floor into an arched ceiling smoothly upholstered in a ribbed pattern, an excellent acoustic damper reminiscent of the padded ceilings in contemporary automobiles, but on a much grander scale. The result was an open, streamlined look that enhanced the Clippers' relatively large interior dimensions.

Norman Bel Geddes's proposed Airliner No. 4 was an ocean liner that could fly. It was designed to offer its 450 wealthy passengers all the amenities they would expect when travelling first class by ship. Although it remained a fantasy, it provided many cues for the designers of today's super-jumbo airliners.

The M-130's 'bridge' accommodated a crew of ten, including a commanding officer, two pilots, navigator, flight engineer, radio operator and relief crew, who slept in bunks located in the plane's nose. From the cockpit, a spiral stair led to the passenger deck, arranged around an off-centre aisle with a single seat to one side and, across the aisle, three abreast seated on thickly upholstered sofas with light tubular steel frames. These sofas had high, straight backs, as in a contemporary European railway compartment, and substantial padded arms, firmly dividing the individual passenger seats. At night the arms could be removed and the sofas converted into sleeping berths, curtained off from the aisle for privacy. In this way the plane accommodated up to 40 passengers on overnight voyages. On day flights it carried 74 seated passengers and still allowed plenty of room to move about and opportunities for general conviviality, much of it involving cocktails and cigarettes.

In Britain, the interior designer Brian O'Rorke worked in a similar vein with the cabins of the Imperial Airways Empire flying boats, first launched in 1936. These vessels, designated S.23, were designed by Arthur Gouge for the manufacturers Short Brothers. They were fast (200 mph) and handsome ships with accommodation on two levels, the upper deck for the cockpit, radio room and crew quarters, the lower deck with a full galley (where meals were prepared from scratch), washrooms, baggage and cargo holds, and two main passenger cabins, furnished with Pullman-style seats, convertible to sleeper berths. The rear 'Promenade' cabin provided a generous aisle along the windows for viewing the scenery below and offered plenty of space for leg stretching during long journeys to the far corners of the British Empire (see p. 8).

All upper sleeping berths were provided with a small window, enabling those resting in their beds to enjoy night views. Every seat had

Pan American Airways operated one of the world's most successful fleets of flying boats during the 1930s. The Sikorsky S40 (top and bottom right) was the first of the great Clipper fleet that served the Caribbean. It accommodated 38 passengers in great comfort and offered elaborate meals prepared on board. The Consolidated Commodore flying boat (bottom left) operated between the USA and South America. Inside, it resembled a Pullman car.

its own hot-and-cold air regulator, a call button and a personal reading light. The dark green wall coverings, trimmed with slender stainless-steel beads, gave the cabins a restful atmosphere enhanced by insulation that deadened the roar of the engines and reduced vibration. Like Bel Geddes, O'Rorke used the visual elements of streamlining to lend the cabins an air of high modernity, with a sense of luxury conveyed through the formal balance of elements and the fineness of detailing.

In this elegant environment, the hospitality was prodigious. One former steward recalled, with pleasure and pride, the Empire Flying boat on which he worked in the 1930s:

[From Waterloo Station] a special Pullman parlour car right to dock-side at Southampton – aboard a little motor ship out into the harbour and along side a beautiful Imperial Airways flying boat, which rode the boarding rain as easily and majestically as a sea gull . . . There were no more than a dozen seats up front. We had orders that every passenger must be addressed by name . . . At lunch or dinner, or at high tea, our attitude was always, 'A bit more of the lobster, Sir?'[31]

Such aircraft succeeded in transporting many thousands of passengers all over the world during the 1930s and '40s. The last and largest of the pre-Second World War flying boats was Pan American's Boeing 314, designated *Yankee Clipper*, for the first transatlantic mail-passenger route, inaugurated in 1939 just months before the outbreak of war in Europe. As on the M-130, this vessel's cabins were arranged as convertible staterooms, the four large cabin spaces separated by curtains at

*The galley of the Scylla.*
*(Weekly Illustrated)*

night. During the day passengers could relax in these cabins, furnished as living rooms or configured for formal dining, an appropriate setting for the elaborate multi-course meals that were prepared onboard and served slowly by the stewards to occupy the time on long flights. The almost Spartan decor was a far cry from the extravagance of the earlier Do x interiors. Nevertheless, the 314 was the last word in comfort for international flying by heavier-than-air craft before the Second World War.

## Cathedrals of the Sky

Despite the grandeur of the flying boats, during the early 1930s unquestionably the most luxurious way to fly was in the German airship *Graf Zeppelin*, which made its first commercial flight in 1928. During the next year, this enormous vessel travelled to the United States, all over Europe, to the Middle East and finally completed a round-the-world cruise, carrying twenty passengers and twice as many crew members. Between 1928 and 1937 *Graf Zeppelin*, along with the newer and larger *Hindenburg*, provided the highest standard of comfort and service offered to commercial air travellers in the first hundred years of aviation.

Following the First World War, the Germans were prohibited for several years from reviving their airship industry.[32] Count von Zeppelin had died in 1917, but his work was carried on by a talented journalist and airship pilot, Hugo Eckener, who tirelessly promoted the peaceful uses of zeppelins, particularly in America, where he eventually gained the support of the powerful newspaper publisher William Randolph Hearst and President Calvin Coolidge, enabling him to construct and deploy a new generation of lighter-than-air vessels to be used for a variety of scientific and humanitarian purposes before entering scheduled service.

Eckener's chief priority was safety, yet he also designed the *Graf Zeppelin* to recapture the glory of the earlier DELAG passenger zeppelins and to surpass all contemporary aeronautical competition for comfort and style. The ship was a symbol of the new German republic, its advanced technology and culture. It accommodated twenty passengers in ten two-berth cabins on its upper deck, and it offered them a cruise exceeding in interest any other means of transport at the time. The style of the interiors was domestic and decorative, with inlaid dark veneers panelling the walls, decorative curtains over the large arched windows and classically styled furniture covered in strongly patterned Art Deco upholstery. The following excerpts from the recollections of Alicia Momsen Miller, who travelled on the ship as a child of eight with her family from Rio de Janeiro to visit the Chicago World's Fair in 1933, demonstrate how astounding all aspects of a zeppelin flight must have seemed:

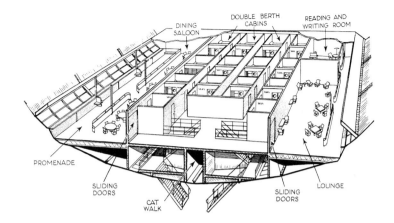

The German airship *Hindenburg* provided a standard of comfort and luxury never matched by a commercial aircraft. Its Promenade Deck, private cabins, Dining Salon, Smoking Room and Library were the ultimate in comfortable long-distance air travel up to the present day.

my father had to convince my mother to travel by air ... 'I want a close up look first!' They drove to the airfield ... There was the huge airship, tied to the ground. It was a very windy day, and its outer covering was shivering. The fabric looked like you could poke a hole through it with your finger. She was horrified, deciding never to trust her children in such a thing. But my father insisted they look at the accommodations in the gondola, and they ascended the short sturdy ladder.

'What a surprise!' my mother recalled. The large living room with its big windows had a number of attractive chairs and tables, and down the hall were wonderful roomy double staterooms. She felt the mattresses, and found them comfortable. The wide bunks were made up with linen sheets and warm fluffy blankets. 'If anything happens, at least we'll all be together,' she said.

Thus, the ship sold the service, its interiors helping to allay the fears of would-be passengers. But it was on departure that the real excitement began, especially for those who had not yet flown in any type of air transport:

we cast off at exactly 6.30 a.m. on this clear and cloudless morning, perfect for air travel. As we slowly rose above the field we could hear the cheers of the crowd fade away and excitedly watched tiny houses, green fields, and purple mountains slip away too, the first of thousands of views.

Miller also described the interior details of the ship:

Inside the comfortable living room (which served as a dining room for meals) of the gondola, the windows were curtained, and red-flowered carpet covered the floor. We could hear a faint humming sound from the motors outside . . . The control, chart, radio rooms and galley were at the front of the gondola. Next came the living room, from which a hall led to ten staterooms, five on each side. The staterooms, decorated with flowered wall paper, had little closets, end tables and sofa beds which turned into double-decker bunks . . . The waiters changed the living room into a dining room by putting white tablecloths and flowers on the tables, setting them with linen napkins, crystal glasses, and china plates edged in cobalt blue and gold.

She added: 'No one on board was allowed to smoke because of the flammable hydrogen gas which kept us afloat. The smokers chewed gum instead.' The eight-year-old Alicia had not, apparently, been taken to the lower deck, where the cocktail bar and smoking room were located. In the latter, a steward was in attendance at all times to ensure that no passenger left the room with a lighted cigarette. He would also light the passenger's pipe or cigar with his own match, while an electrical device was available for passengers to light their own cigarettes.

Miller continued: 'We went low enough to see through the open windows of some houses, and people in the streets and houses waved to us with their towels and cloths.' The zeppelin could hover, virtually still, above the earth, as it did when landing, offering those aboard and those on the ground a rare form of interaction, similar to, but perhaps even more exciting than, the exchanges between passengers on the deck of a docking ocean liner and their friends and loved ones on the pier. Despite such charming opportunities and the domestic civilities of the gondola, it was the gargantuan engineering of the ship's architecture that most impressed some passengers. The Momsen family was invited to see the inside of the hull of the ship:

We were inside the silver cigar! Above us were wires and girders and big bags of hydrogen gas. Below were canvas water containers and fuel tanks. It was very quiet, like being in a huge warehouse. We walked along a spindly metal catwalk that went from one end of the ship to the other. A flap of skin over one of the motors was open, and I can still remember the strange feeling in my stomach as I held tightly to the handrails and looked straight down at the tree tops below. A member of the crew, off duty, took his accordion to the nose of the ship and played for us there. It was exciting, but I didn't feel very secure until I got back to the solid-feeling gondola.[33]

The inside of the hull of a giant zeppelin was certainly one of the epic spaces of the twentieth century, indeed of all time; and the largest interior of all belonged to the younger sibling of the *Graf Zeppelin*, the *Hindenburg*, planned by Eckener but completed by his Nazi successors, who seized control of commercial airship transport after Hitler's accession to power in 1933. *Hindenburg* joined the *Graf* in service from 1936 and made ten transatlantic round-trip voyages before crashing in flames at Lakehurst, New Jersey, in 1937, killing 35 of the 97 people aboard. Despite various credible theories, the cause remains an unsolved mystery.

Not only was *Hindenburg* larger, it was also more luxurious than the *Graf*, with 25 double staterooms, reached from the entrance on the lower deck via a grand staircase. There was a library, smoking room, a 200-foot promenade deck, a 50-foot-long dining room, a lounge equipped with a pigskin-covered aluminium piano and three bars, open at all hours. Due to weight restrictions, furniture throughout the ship was made of tubular steel or aluminium, executed by Fritz Breuhaus de Groot in a Bauhaus or International Modern style, in spite of Nazi hostility to modernism in architecture and design. Modernism, it seems, was acceptable in cars and aircraft, but not in building design, fine art or domestic furnishings.

There were also traces of the Art Deco, such as a prominent wall mural depicting a stylized map of the lands over which the ship would pass, and colourful, semi-abstract paintings of travel scenes. The German zeppelins were without rivals in their time, or at any time since, in providing a unique flying and travel experience that can be recaptured only by the reports of those who experienced it. Among the many accounts and recollections of the *Hindenburg*, one of the most evocative is that left by the novelist and social philosopher Arthur Koestler, who flew on the ship in 1936 and wrote:

> The passenger gondola, as it was so poetically called, was attached below the hull, near the bow. It contained the bridge and the passenger cabins, which were furnished with all the luxury of a modern ocean-going liner. At the end of the gondola was a door that was usually kept closed. From it a dark corridor led to a land of mystery, into the belly of the whale. And it really was the most fantastic place that you could possibly think of.
>
> Darkness reigned in the whole of the interior, whose dimensions were greater than those of a cathedral, except that they were elongated in the shape of a cigar, it smelt like bitter almonds and

there was a dull grinding sound as though invisible bats were lazily flapping their wings. The smell came from the hydrogen cyanide that was mixed with the fuel and the noise from the balloon envelopes, the floating whale's most important organs, its air-bladder as it were, as it was their enormous buoyancy that carried the gigantic structure through the skies. These balloons were arranged in two rows along the whole of the interior and hung down like enormous pears, each about 15 metres in diameter. Under normal air pressure, when the airship was on the ground, they hung slack, with loose, wrinkled skin, like old witches' breasts; but when the airship climbed into thinner air they became tight and firm and filled the belly of the whale with a noise like a thousand whips being cracked. Between the rows of balloons, suspended freely in the air and in complete darkness a dangerously narrow catwalk hung at a height of 20 metres; the gas cells were serviced from here, by the light of miners' lamps, because of the enormous danger of explosion.

Although it was permissible to use this catwalk only when accompanied by a ship's officer I often wandered about on it, which was always a profoundly exciting experience. In the surrounding darkness, amidst the soporifically bitter almond smell the trembling of the body of the ship and the grinding and banging of the balloon envelopes was overwhelming. I was surrounded by a framework labyrinth of girders and spars, cables and ribs, a jungle of steel and aluminum, while only a few hundred metres below extended the nameless white desert of the deep North.[34]

## Flying Fortresses

By the outbreak of war in Europe in 1939, the airship era had ended. Although it was not generally recognized in the aviation industry, the flying boat had also had its day. Because of the rapid construction around the world of airfields to serve the war effort of both Allied and Axis nations, the future of aviation securely belonged to land-based aircraft. Bombers and other military planes designed in the 1930s represented a major advance in size, speed and payload over those of the previous generation. Most of them were built entirely of metal, although some of the most successful aircraft of the period, such as the British de Havilland Mosquito fighter-bomber, were built of plywood. The new planes were equipped with retractable landing gears, more advanced navigational instruments and internal radio communications systems to link the crew.

Perhaps the greatest workhorse of the war and the heavy bomber built in the largest numbers, thanks to the vast production capabilities of

North American manufacturing industries, was the heavily armoured Boeing B-17, the Flying Fortress. When it entered service in 1939, the B-17 was the fastest bomber in the world and was capable of the highest altitudes. It also boasted a well-designed interior.

One retired RAF pilot observed: 'the B-17 has a functional but welcoming cockpit. There is nothing plush or fancy about it, but all seems purposeful, and there is a lot of light, the generous provision of transparent surface lends the aircraft a conservatory air.' The raised cockpit of this large plane gave the ship an elegant profile somewhat like the latter-day Boeing 747 airliner. Internally, it provided great all-around visibility and its split Veed windshield and domed rooftop suggested the passenger compartment of a big streamlined American car of the 1930s. Yet the cockpit's very lightness and transparency may have left combat crews feeling somewhat exposed and vulnerable in battle.[35]

The ergonomics of military aircraft are often incomprehensible to a civilian as are the rituals they generate. The very size of a B-17 might suggest that it could be accessed much like an airliner of the period via a set of steps leading through a door in the side of the fuselage, and indeed there was such a door. However, the pilots entered by a different route, as described below:

The cockpit of the B-17 is a pleasant place to work, but accessing it is another matter. Honour demands that pilots enter through the forward hatch, which is only some two feet square and six feet off the

The Flying Fortress did not always live up to its name. The nose gunner occupied one of the most exposed positions of any military air crewman. The cockpit was bright and spacious and bore some resemblance to a streamlined American car of the late 1930s.

ground. The Gregory Peck entry involves grasping the outer edge of the hole with both hands, lifting and swinging both legs together through the opening, and then twisting them down the fuselage in the process, so that with a final heave and squirm the body is deposited on the cabin floor, quivering from exertion.[36]

As the first highly successful American bomber, the Flying Fortress became the stuff of legends. One particular plane was immortalized, first in a film documentary, *The Memphis Belle: A Story of a Flying Fortress*, directed by William Wyler in 1943 and produced by the film unit of the US Army Air Corps and Paramount Pictures. This film documents the *Memphis Belle*'s last mission, a bombing raid on German submarine pens. It was filmed with hand-held cameras inside the plane to give a realistic impression of the crew's experience on such a mission and also providing a detailed view of all areas of the plane's interior.[37]

A Hollywood film of 1990, *Memphis Belle*, directed by Michael Caton-Jones, fictionalized the experience documented in the earlier film and also provided an even more vivid, if romanticized impression, this time in colour, of the inside of a B-17. The film showed how the crew functioned as a unit within the separate parts of the plane's interior. They were linked by intercom to each other and, via radio, to the crews of other planes in their formation, which was a tightly organized fighting unit that had to carefully coordinate the simultaneous release of their bombs to achieve the objectives of their missions.

High-level bombing accuracy was dependent both on the size and precision of the formation and on the newly invented Norden bombsight, a gyroscopically stabilized instrument, which was fed information by the bombardier through a state-of-the-art electromechanical calculating device. The Norden bombsight transformed high-altitude bombing from a crude, seat-of-the-pants exercise into a complex art. Thus, the view of the world from the aircraft was rapidly becoming more reliant on sophisticated technological assistance. The distinction between 'sighting' and 'seeing' could not be more dramatically revealed than in the situation of the bombardier, squinting through a Norden bombsight, while seated behind the transparent Plexiglas nose-cone of the B-17, with the extraordinary, panoramic view that position offered.

The European plane most similar to the Flying Fortress was the fast and robust, British-built Avro Lancaster bomber, which was most celebrated as the transport of the RAF's 617 Squadron (aka Dam Busters) who destroyed several strategically important dams in the Ruhr Valley with highly innovative 'bouncing' bombs. The Lancaster also became notorious as the plane that carpet-bombed Dresden in 1945, causing a fire-storm that

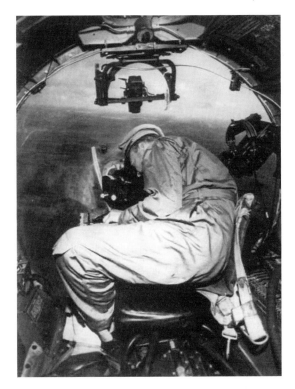

killed more than 50,000 civilians, including many refugees. 'It's distance and blindness that let you do these things', recalled Hans Boles, gunner on a Lancaster bomber operating at 20,000 feet during the night raids that devastated German cities in the latter stages of the war. The journalist and military historian Max Hastings also commented on the ferocity of Second World War bombing of civilian targets:, 'this was a far cry from the "foraging" missions of World War One'. His comment recalled the dire speculations on the ghastly potential of aerial warfare published by H. G. Wells at the beginning of the last century – aircraft 'dripping death'![38]

Despite its dubious deployment and questionable accomplishments, the Lancaster was regarded with deep affection by those who flew it. Flight Lieutenant H. T. Goodman declared that the

The clear Plexiglas nose of the B-17 was the location of the Norden bombsight, illustrating the growing contrast between seeing and sighting in modern aviation.

Lancaster cockpit was beautiful . . . warm and comfortable, and the view through the canopy was very good. There was even a bulge on the side of the canopy where you could see down and towards the back. The Lancaster had no bad habits . . . [and] practically flew itself.[39]

Another former pilot compared the thrilling roar of the Lancaster's four Merlin engines to the rich and complex sound of a full symphony orchestra.[40]

Of the many other Allied and Axis planes of the 1940s that achieved distinction, one of the most successful and visually extraordinary was the Lockheed P-38 Lightning, which first flew in early 1939. This twin-engine, high-altitude interceptor featured a unique overall configuration, with a short, teardrop-shaped cockpit faired into the continuous wing and flanked by two long booms, containing the engines, and swept back into a striking twin-tail assembly. The Lightning was capable of 400 mph and an altitude of 25,000 feet, and it also had the longest range of any fighter operating in the Pacific, where the distances between base and target were the greatest in the war. Due to its long range, the P-38 achieved its most outstanding success by destroying, in 1943, through stealth and surprise, the plane carrying Isoroku Yamamoto, admiral of the Japanese naval fleet.

Like most American fighters, and in contrast with the tight cockpits of typical European fighter planes, the Lightning was big inside. From the pilot's perspective, the P-38 was something quite special, as was its most famous pilot, Antoine de Saint-Exupéry. The aviation journalist Jeffrey Ethell described his first experience in a Lightning, sitting under its domed Plexiglas canopy:

> Once I was settled in the cockpit, I was taken with the vast expanse of airplane around me. Sitting deep within the centre gondola and wing, I quickly got the impression of being buried within the machine; this would intensify in flight. The cockpit is just about perfect in size, not too small, not too large and very comfortable.[41]

Not only did the Lightning fit its pilot, it provided a level of ergonomic efficiency unmatched in comparable aircraft of its generation. Ethell described these qualities:

> The most obvious difference from other wartime fighters . . . is the dual pistol grip control wheel. Putting both hands on this brings a sense of complete authority. I can see why it was so easy to haul the aircraft into tight turns; both biceps are working. The ergonomics of the wheel are also years ahead of their time: the grips are canted inward to the exact position of one's hands when they're relaxed and held out in front of you.[42]

The sensation described above was what the Vice-President of General Motors, Harley Earl, was trying to communicate to the drivers of GM cars when he aped the twin boom design of the Lightning in the shapes of his Cadillacs, Buicks and Oldsmobiles in the late 1940s and throughout the 1950s. The external appearance of these cars derived their tail fins and nose-cone-shaped front bumpers directly from the P-38. More important to creating the feeling of being in the plane, however, was the roof and window treatments of these cars, which imitated the Lightning's canopy and the spaciousness of its cockpit, with plenty of 'ship' around the driver. By the mid-1950s GM cars featured compound-curved, wraparound windshields, which had initially challenged the production capabilities of the glass-makers Libbey-Owens-Ford. Designers of the greenhouses above driver and passengers strove relentlessly to match the glassiness of the P-38 and later jet fighter planes in order to create a fantasy element in post-war American cars. This was parodied in the comedy film *The Secret Life of Walter Mitty* (1947), based on James Thurber's short story, in which the daydreaming hero, driving his car, imagines that he is actually a war ace at the controls of a military fighter plane. Thus, Earl's cars were meant not just to look like the P-38, but also to feel like it.

Unlike the glamorous Lightning, the C-47 cargo plane (aka Dakota, Gooney Bird, Skytrain), a military derivative of the twin-engine Douglas DC-3 passenger aircraft of 1935, was a stolid but ultimately heroic workhorse during the Second World War, in the period of post-war reconstruction, and throughout the conflict in Korea, the Cold War and the war in Vietnam. More than 10,000 were produced to transport military cargo and personnel around the world, and its role in the Second World War was immortalized in films and TV series such as *The Longest Day* (1962) and *Band of Brothers* (2001). The C-47 was fitted out in a variety of forms: as a basic cargo plane, as a trainer and for reconnaissance duties; and some had plush, military VIP interiors with high-quality sound insulation, walnut panelling, thickly upholstered, leather-covered armchairs, sleeper berths and meeting tables. The primary purpose of the C-47, however, was to move troops in large numbers.

In the configuration of a personnel carrier, the austere tubular interior of the fuselage was arranged with anatomically contoured benches, seating 21, running along the external cabin walls the length of the passenger compartment. The floor was bare plywood and the interior sprayed in Federal standard 'interior green' throughout. Seat belts were provided as a concession to safety that may seem ironic when so many of the passengers were about to jump out of the plane through anti-aircraft fire and parachute into battle. The lavatory in the tail of the plane consisted of a bucket and paper holder.

In front of the passenger cabin there was a radio operator's compartment, a navigator's position with a large wooden map-reading table and the pilots' compartment, seating two. The deeply Vee'd windshield provided a panoramic view extended by two Plexiglas side windows, which could slide open to ease communication with ground crew. There was a wide array of instruments, arranged on a flat panel painted matt black to avoid glare, and two contoured bucket seats upholstered in a standard apple green.

In the cockpits of these planes, pilots endured some of the harshest flying conditions ever experienced: anti-aircraft fire, tight formations in crowded airspace, inadequate runway facilities, round-the-clock flying in all weather, and planes overburdened by heavy loads and punishing schedules required to prosecute war and maintain peace.[43] The provision of 'Rebecca' and 'Eureka' navigation systems for blind flying in the mid-

1940s helped pilots to cope with bad weather and the night flying required, for example, by the Berlin Airlift (June 1948–May 1949) during the Soviet blockade of the western sector of the German capital. But these advances did not eliminate the dangers faced by c-47 crews and passengers. These were the heavy trucks of the air, and their interiors reflected the austerity of the times and the tasks they had to carry out.

The dark, austere and foreboding passenger cabin of the c-47 transport plane contrasts with its bright cockpit. This great warhorse saw service in every part of the world, carrying millions of military personnel for more than fifty years.

Their most dramatic service was performed on 6 June 1944, when vast formations of c-47s carried more than 40,000 paratroopers into combat over northern France on D-Day. Parachute drops were made from altitudes as low as 600 feet, at 90 mph, in a sky filled with ferocious anti-aircraft fire. Some paratroopers preferred the dangers of the jump to the hazardous journey inside the c-47 – and nothing in the interior design of these planes could have reassured those soldiers, looking out through the Plexiglas windows at the Normandy coast, when the jump light was activated and the fuselage door opened early that blustery summer morning.

Other pilots of the 1940s were testing their physical and emotional stamina against the severities of flight in different ways. Test pilots, whose profession is to endure the most challenging effects of air travel, were testing not only the physical structure of the newest planes, but also the limits of human physical endurance. Charles Lindbergh recounted his experience testing a Republic P-47 'Thunderbolt' fighter plane over the Ford Company's River Rouge assembly plant in Detroit, where the plane was built, in 1943:

I shove the stick forward. The earth slants upward and the dive begins . . . 35,000 feet . . . 34,000 . . . my cockpit roars through the air . . . the earth fades out . . . the instrument dials darken . . . breath's thin; lungs empty – I'm blacking out – losing sight . . . I push the nose down farther . . . faster . . . 33,000 . . . 30,000 . . . the dials become meaningless . . . down . . . down . . . I am dimly aware of a great shriek, as though a steam whistle were blowing near my ears . . . it's oxygen I need . . . I'm blind . . . there are no more seconds left – it's a razor edge – a race between decreasing consciousness and increasing density of air . . . 17,000 . . . 16,000 . . . 15,000 . . . a white needle moves over white figures . . . it's the altimeter – I can see – I'm aware of the cockpit, the plane, the earth and sky . . . I've already begun to pull out of the dive – the air in my lungs has substance. Perception floods through nerve and tissue. How clear the sky is above me, how wonderful the earth below.[44]

In the late twentieth and early twenty-first centuries, electronics and automation changed the pilot's life aloft considerably. The archetype of the new direction in aircraft design is the Northrop Aviation B-2 'Stealth Bomber'. The B-2 is a heavy, long-range bomber with 'low-observable' qualities, derived from its 'flying wing' configuration and angular surface geometry that enable it to avoid detection by radar and other defensive systems.[45] It can operate at an altitude of 50,000 feet and has a range of more than 6,000 miles. Since its inauguration in 1993, it has been deployed in bombing raids over Serbia, Afghanistan and Iraq. It is the modern equivalent of the Boeing B-17 or the Avro Lancaster bombers.

The crew of two required to fly this big plane is accommodated in a cockpit contained in a bulge midway across the wing's leading edge. Like the pilots of the B-17, the B-2's crew enter the spacious flight deck through a hatch in the belly of the plane. Once inside, they have an extensive view, forward and above, through the large rectangular windows that wrap around the seating area. While the pilot flies the plane, using a fighter-style control stick rather than a steering wheel, the commander navigates and deploys the weapons, which include GPS-guided 'smart' bombs and up to sixteen nuclear weapons. The crew sit in the McDonnell-Douglas 'Aces II' rocket-catapult ejection seat, an electronically operated, aluminium monocoque structure ergonomically tailored to long flying times. The ship can carry a toilet and there is space for a foldable bed, but no further amenities are provided as standard equipment.

In addition to its stealth geometry, it is the B-2's 'glass' cockpit that marks its interior as modern. The dashboard presents an array of eight multi-functional video display screens that comprise the electronic flight instrumentation system

The Northrop B2 Stealth Bomber is easily recognized by its sharp geometry, echoed in the rectilinear formality of its 'glass' cockpit, which features instrumentation familiar to members of the Nintendo generation.

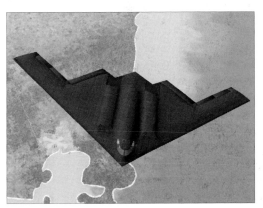

(EFIS). Using this system, the pilot can select the appropriate display of data for the specific task to be undertaken, a complicated process, but one that comes naturally to a generation of pilots raised on video and computer games.[46]

## Seat and Tray

In the early days of commercial aviation, airliner interiors were designed to resemble yachts, limousines, ocean liners or Pullman carriages. But by the mid-1930s a specific set of parameters were evolving that would demand the development of interior layouts and details specific to the airliner. IATA (International Air Transport Association) had set out specifications for cabin and seat design that had to be met by carriers in order for their planes to receive certification; underlying this was the assumption that engineering and safety would determine all aspects of design in conjunction with, or in conflict with, the economic impera-tives of the aircraft manufacturers and the airlines. From the passengers' perspective, however, comfort remained among their highest priorities, and the airlines competed to offer new solutions to the provision of the most comfortable interiors, and in particular the best airline seat.

In a lecture to the British Institute of Transport in 1936, the speaker noted the need for airlines to provide the passenger with

> space, good ventilation and heating, agreeable food, an attentive steward and, above all, his chair. By sea an outside room with a bath is demanded, but by air the passenger's time is spent chiefly seated in a chair and, depending upon his mood, that chair must enable him to sit upright when taking his meals, or to loll back in luxurious com-fort to read, and in it he must be able to recline or doze or sleep. All this must his chair fulfil, and that independently of his fellow passen-ger, who may require his chair to do quite the opposite and at the same time. Imperial Airways has recently completed a chair of its own invention which is fitted as standard to the new flying boats and landplanes. Its comfort and adjustability show a marked improve-ment on anything of its kind yet in use; moreover, its weight is only 18 lbs – a vital factor from the aircraft operator's point of view.[47]

Airlines, in collaboration with specialist furniture companies such as Dryad in Britain, developed more ergonomically considered seats that the passenger could adjust to change position during long-distance flights. Such seats were installed on British Empire flying boats and the Imperial Airways Armstrong Whitworth Ensign, both with interiors designed by Brian O'Rorke. These planes provided excellent testing grounds for innovative seating to be enjoyed by the privileged few. With lightweight aluminium frames and removable padded covers, the seats had adjustable backs and hygienically covered wings for resting the head. Their cushioned armrests were positioned above the exposed tubular framework, giving the chairs a look of fashionable modernity.

Yet in the 1930s it was the work of the Douglas Aircraft Company in the USA that helped to bring comfort to millions of airline travellers in the design of a series of highly successful aircraft introduced between 1932 and 1935, the DC-1, DC-2 and DC-3. Designed by Donald Douglas and Jack Northrop, these civilian prototypes of the C-47 transport were twin-engine monoplanes of all-metal construction, employing the most advanced streamlining and sophisticated production technologies to provide the fastest, strongest, most comfortable and most efficient planes in the sky. The final version, the DC-3, was used by most major airlines in the world. By 1939 it had become the first passenger plane to achieve profitability without government mail subsidies and was carrying more than 90 per cent of the world's passenger traffic. This was truly the Model T Ford of the air.

The Douglas designers paid careful attention to passenger comfort in their designs for the interiors of the DC-3, correcting many of the faults evident in previous airliners, such as the Ford Tri-motor. Because of its advanced, monocoque construction, no structural members intruded into the tubular cabin, giving the designers a free hand in the division of the tubular space and permitting unimpeded circulation throughout the ship. The cabin was extensively soundproofed, keeping interior noise levels low enough to allow conversation in a normal tone of voice. While there was a Pullman-style convertible sleeper version, the DST, most were outfitted in daytime configuration. Every seating position had an individually controllable fresh-air vent, while cabin air was recirculated once a minute; and a steam boiler, heated by the engine exhaust, kept the cabin comfortably warm.

While many planes and flying boats of the 1930s were furnished with plush, rigid, two- or three-person couches, the DC-3 seated up to 28 passengers in individual, forward-facing, high-backed reclining seats with tubular aluminium frames mounted on rubber pads to suppress vibration. Seats were in pairs on one side of the aisle and single on the other. These chairs represented a simple development from the adjustable railroad seats, patented in the third quarter of the nineteenth century and described by Siegfried Giedion in *Mechanization Takes Command*. But the airline seats were considerably lighter because of the new materials used, including magnesium for parts of the frame and foam rubbers of

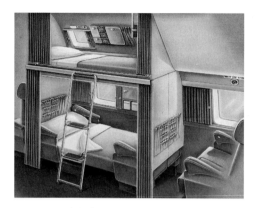

varying density for the padding. They also featured more sophisticated spring and cog mechanisms used for adjustability. Giedion applauded the DC-3's chairs in relation to the advance they made in personal comfort for air travellers and the technology of their construction. But he was characteristically critical of their relationship with contemporary furniture style. He wrote:

> Although well-designed from the point of view of light materials and in the best tradition of patent furniture, [the Douglas adjustable aircraft seat] suggests a trend to artificially heavy appearance – the

The smoothly streamlined Douglas DC-3 aircraft, used by nearly every major airline in the world, were furnished in a wide variety of configurations, including a sleeper version and day models with seating for two, three or four abreast.

outcome, it would seem, of 'streamlining', which perpetuates the showiness of nineteenth-century ruling taste in so many areas.[48]

The chairs designed by the Douglas Aircraft Company did look very similar to chairs used in Pullman Parlor Cars in the 1870s, with the exception that modern airline seats showed evidence of the radius curves so typical of streamlining. They also featured lateral head supports, derived from traditional wing chairs, narrowed and enlarged to cradle the head of a sleeping passenger more effectively, and linked by a bolster to support the neck. Although engineered for lightness, they were styled in line with contemporary popular taste, and their apparent heaviness may also have been a concession to the reassuring impression of comfort, security and familiarity that the designers of aircraft interiors had been attempting to convey in various ways since the start of commercial aviation.

Aluminium- and magnesium-framed seats with an array of innovative padding devices, including inflatable cushions (also intended as flotation devices on flying boats), kapok and foam rubber, were being produced for aircraft manufacturers and airlines in Europe and North America from the early 1920s by new, specialist furniture manufacturers and by established furniture companies. For Thonet, Anton Lorenz and Hans Luckhardt carried out extensive research and designed a series of elegant, fully reclining, cantilevered, aluminium-framed airline seats in the 1930s. In the US the Heywood-Wakefield furniture company also engaged in extensive studies of ergonomics to produce more efficient and comfortable airline chairs.

Among those who set up firms primarily to cater to the aircraft industry, the designer Warren McArthur of California and New York is notable for his innovative use of light metals and for more than 40 different seat designs he produced between 1935 and 1948. The Warren McArthur Company also manufactured an overwhelming percentage of the pilot, crew and passenger seats installed in American military aircraft during the Second World War.

Although it greatly broadened the flying public, the DC-3 was still a luxury liner, used primarily by the well-off. Yet it was an important plane because it served so many different purposes, endured so long and broke the tradition of aircraft interior design

Warren McArthur's airline chairs bore strong resemblance to his domestic furniture. This modern furniture style developed simultaneously in both transport and domestic design during the 1930s.

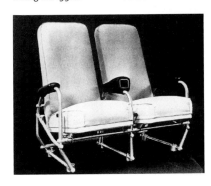

imitating other forms of transport, such as trains and ships. It also achieved a change of mind among passengers, away from expecting an experience akin to that on the *Queen Mary* or the Atchison, Topeka and Santa Fe Railways' *Super Chief,* towards a sense of air-mindedness that accepted the regime of sitting in an assigned seat for the duration of the flight.

Possibly the single greatest technical innovation that changed the nature of passenger air travel was the advent of the pressurized cabin, which was perfected in the later 1930s but not employed in a new generation of mass-produced airliners until after the war. A derivative of the B-17 Flying Fortress, the Boeing 307 Stratoliner was the first commercial aircraft to have a pressurized cabin and an operational altitude of over 25,000 feet, enabling the plane to cruise above the weather and therefore to provide its passengers with a level of flying comfort previously unknown. This plane was produced in small numbers in 1940 and delivered to Pan Am and TWA, but the latter were requisitioned by the military soon after commissioning.

The fuselage of this tail-dragger was circular in section, like the DC-3, but had adequate space for the provision of a row of four cabins, running up the starboard side, each furnished with two rows of triple seats, facing each other. These could be converted to twin-berth sleeping compartments, separated by an aisle from a line of nine individual armchairs. The maximum capacity was 38 passengers in a non-sleeper version. These had two pairs of seats per row. The Stratoliner was a transitional aircraft that could offer its passengers the traditional luxury of a sleeping compartment, or such advanced amenities as in-flight movies

Inside the organically shaped fuselage of the Lockheed Constellation, first-class passengers enjoyed a domestically furnished living room capable of turning a flight of several hours into an impromptu cocktail party.

(pressurization quieted the cabin sufficiently to make sound projection practical), in a controlled internal atmosphere that successfully defied the extremes of the natural elements.

Also designed before the war but not entering commercial service until 1946, the Lockheed Constellation was the first pressurized, high-altitude, land-based plane to be produced in large numbers. The original 'Connie' and its two enlarged offspring, the Super Constellation and the Starliner, introduced in 1950 and 1957 respectively, became the most successful piston-engine airliners ever produced, conducting the first non-stop transatlantic services as well as direct, long-distance flights all over the world with all the major airlines. Its sleek, organically contoured fuselage and triple-tail boom, designed by Kelly Johnson and the engineer Hall Hibbard,[49] made the 'Connie' one of the most distinctive and elegant airliners ever built.

Its tall tricycle landing gear demanded an equally tall flight of stairs to enter and exit, providing spectacular photo opportunities for the celebrities who flew the plane throughout the 1950s. As Eisenhower's presidential plane, the Constellation served as a global symbol of American engineering and design excellence, and represented the highest values in modern transportation between 1946 and 1959, when it ruled the air. It became a widely admired icon of the technological bio-morphism popular in all fields of art and design between the 1930s and the 1960s – consider the shapes of Eero Saarinen's 'Tulip' chairs or Alexander Calder's mobiles. The aviation historian Roger Bilstein expressed the romantic attachment many flyers had to the 'Connie' in his description of its characteristic sound: 'The engines swelled to a deep, melodic rumble like the sustained roll of a battery of tympani.'[50]

The passenger's view, from inside the Constellation and all the later jet airliners, began to lose significance, since little could be seen of the ground from the lower stratosphere, above the clouds. Because there was now less distraction or interest provided by the view, the interiors became a greater focus of attention and had to be even more carefully considered than in previous commercial aircraft. Raymond Loewy designed the interiors of early Constellations, but the later designs by Henry Dreyfuss Associates, under the supervision of William Purcell, had the greatest impact on subsequent aircraft interiors.

The Dreyfuss agency was well prepared to design an interior in which passengers would be restricted to their seats for more extended periods than in the past, since for years they had been conducting extensive research into human measurements and proportions as a basis for highly technical design problems, particularly in relation to transport vehicles. Indeed, they worked for several years with Dr Janet Travell of the Cornell University medical school in studies to establish comfort criteria for aircraft seating. Dreyfuss's office compiled a vast body of information about the average physical dimensions of potential flyers and put them to work as the basis of their plans for the interior of the new plane. Dreyfuss, however, wrote a cautionary note about the idea of 'average' dimensions:

> Particularly important, these dimensions include not only 'Joe and Josephine' – the theoretically average man and woman – but the largest and smallest people likely to use the products we designed . . . A good design must 'fit' not only the theoretical average, but his large and small brothers.[51]

The aim of reassuring timid flyers was made easier by the horizontal floor of the 'Connie', which resulted from the new tricycle landing gear, a departure from the steeply sloping floors of earlier tail-draggers such as the DC-3 and the Stratoliner. The Dreyfuss cabins also emphasized restful comfort. In coach class, an early concession to mass flying, indirect lighting, upholstery, carpets and wall panelling in soft colours and wood veneers created a comforting, domestic ambience. Meanwhile, in first class, the decor was that of an exclusive cocktail lounge. In 1953 the chic Dreyfuss interiors were featured in *Harper's Bazaar* magazine as the background for a line of luggage, cosmetics packaging and clothing in a colour called 'Constellation Blue'.

The 'Connie' was able to offer service that maintained the standards of pre-war airliners, but for a much larger passenger group. This was made possible by technological innovations such as deep-frozen food, introduced about 1950. Now, large numbers of previously prepared meals were brought on board frozen and had only to be heated before serving. All aspects of the plane's interior specification and flight service were, by mid-century, controlled tightly by IATA, leaving only matters of style and detail to the airlines to generate a competitive edge over their rivals.

Although some of the early post-war Constellations were equipped with convertible sleeper berths, the great majority was configured to seat between 44 in the smallest early models and 99 passengers in the latest Starliners.

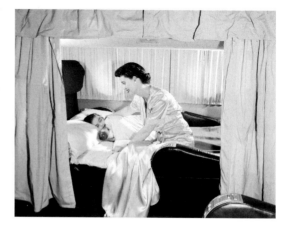

Sleeper seat or convertible bed? The airlines' aspiration to offer high fare-payers a flat bed on long-haul flights has persisted in a variety of forms since the dawn of commercial aviation. The Boeing Stratocruiser made transatlantic flights of more than 15 hours more acceptable to the Pullman generation.

In the earlier of these, just two 'sleeper seats' were positioned at the front of the cabin, offering their fortunate occupants a 45-degree reclining seat back and a 45-degree extending footrest to provide a very comfortable experience in the early 1950s, when the New York to London flight took seventeen to twenty hours, depending on the number of refuelling stops made in Greenland, Ireland or Scotland.

## Jets for Everyone

In 1959 the twenty-year-old Constellation was quickly made obsolete by the introduction of the Boeing 707 jet airliner, an event that marked the start of a new era of air travel. Flying typically above 30,000 feet at speeds of around 550 mph, and with a range of 7,500 miles, the 707 cut travel times for long-distance flights by several hours over its piston-engine predecessors. Because of the shorter duration of many flights, the need for full sleeping accommodations diminished, except as a luxury reserved for a shrinking number of first-class passengers, as pressure increased to pack many more economy passengers onto each flight in order to reduce their fares. The era of mass air travel had finally arrived.

Following the tradition established in the US of hiring consultants to plan the interiors of new aircraft, Boeing hired the interior designer Frank Del Giudice of the Walter Dorwin Teague Agency to design the original layout of the cabins and the seats for the 707. The project began in 1952, using the first full-scale working mock-up of a cabin interior, constructed in the Teague offices. The primary function of this method was to aid in the development of the design and to test its features in simulated operation. But the mock-up also became a sales tool in marketing the plane to the airlines, whose own consultants or in-house furnishing departments would specify the finishing touches.

The 707's first-class cabins employed the usual informal, cocktail lounge banquet seating and other de-luxe flourishes expected by those passengers paying very high fares. Yet in the large economy cabin, Del Giudice's design set a milestone for cool elegance, with three-by-three seating across a wide aisle, highly controllable cabin light-

The first mature jet-space, on board the Boeing 707 (top), was the model for all subsequent economy-class jetliner cabins. Its layout, seat design, material finishes and many of its cabin details remain familiar in the twenty-first century. The Boeing 737 (below) features fully enclosed overhead storage lockers and a dramatically coffered ceiling, which creates an appearance of greater height and width, like the nave of a Renaissance church.

ing, subtle pastel colours, hard wall surfaces and simple clean lines emphasizing the great length of what had become in the 707 a very grand space. The smaller, more closely positioned windows of this plane meant that however the movable rows of seats were spaced, every passenger would get something of a window view.

The 707's economy seat was designed by the Teague Agency in collaboration with the manufacturer, Aerotherm, using the plastic shell technology pioneered in the innovative furniture of Charles Eames and others during the late 1940s and early 1950s. The back of the Aerotherm chair was a crisply sculptured, hard plastic shell with a hinged tray table let into its back. Seat and squab were covered with foam and upholstered in synthetic fabric. The shell was attached to a lightweight metal underframe, incorporating the footrest and the cantilevered arms. The recline mechanism, assisted by a hydraulic cylinder activated by a button, was located in the armrest. The advantages of the new materials and technology were, as usual in all aspects of aircraft design, their savings in weight and bulk.

The use of new materials made a significant difference in both the functionality and the ambience of the cabin. Replacement of soft upholstery by hard plastic finishes was perhaps the most important change. Wall panels and bulkheads were made of rigid PVC plastic with decorative patterns silkscreened or roller-printed onto them. The moulded wall panels now included recessed window reveals and built-in sliding window blinds, creating a continuous surface that eliminated the need for domestic-type curtains. The new plastic panelling was easy to remove in standard sections for repair or replacement; it was highly resistant to damage and easy to clean. Also, such surfaces had an appearance of high quality and durability hitherto not associated with plastic products. They were also warm and pleasant to the touch, and their surfaces responded well to all lighting conditions, from the bright levels of sunlight experienced at high altitude to low levels of artificial lighting used on night flights.

These interiors set the pattern for all subsequent jetliner interiors, although airlines also directed their own interior designers to establish a recognizable decor for their planes. Corin Hughes-Stanton wrote in the 1960s: 'the interiors of aeroplanes are the chief showcase of the airline', and so they must represent the values of the airline, its cultural traditions and its attitude towards the passenger.[52] Yet he also explained that while eye appeal is an important factor in establishing an identity and brand loyalty for an airline, tickets are sold by fare and by the convenience of a timetable, long before the passenger ever sees the cabin. Therefore, comfort and safety are, in effect, the dominant factors in cabin design for 'mass transporters'. He explained: 'seat comfort in aeroplanes is not so much a luxury as a basic necessity because conditions are cramped and passengers cannot move around'. Therefore, modern jetliner 'seats are the most advanced and rationally developed pieces of furniture designed and constructed by modern industry'.[53] This superlative has also been applied to car seats.

Among the carriers who first flew the 707, Pan American Airways made perhaps the most ambitious effort to refine the plane's design as part of a wider corporate identity. The airline engaged the architect Edward Larrabee Barnes to plan a complete overhaul of its corporate image, including the design of its logo, graphics, the livery of planes and ground vehicles, and, most importantly here, the interiors of the 707, whose introduction initiated the entire project. In their design for the 707 cabins, Barnes and his associate, Charles Forberg, finally overturned the lingering assumption that passengers would feel less anxious in an aircraft cabin that resembled the inside of a building or a train.

The Pan Am jet cabins appeared nothing like a hotel lobby or a cocktail lounge, or anything other than what they were, 'the interiors of a finely tooled machine'. The very simple cabins, finished in vinyl laminates and PVC panels, used the revised corporate colour scheme of sky-blue and white for interior surfaces that appeared smooth and hygienic, light-absorbent and tactile, producing a calming atmosphere. The textile designer Jack Lenor Larsen was engaged to create for all Pan Am planes specially woven fabrics in the new 'recognition' colours of blue, grey and black, with green accents in tweed or striped patterns.

The Barnes office worked with the seat manufacturers, Aerotherm, to develop a more refined seat within the tight regulations set down by IATA. A jet airliner seat in the 1960s had to withstand acceleration and deceleration of 9 g; it had to be light (around 25 lb or 11.4 kg per seat); it had to have impact-absorbing elements to reduce injuries in an accident; it needed to be easily serviced and moved; and it incorporated a hydraulic recline mechanism and control, removable seat covers fastened with Velcro, a footrest, a folding tray, literature pocket, seat belt, ashtray and call button. Through their collaboration with Aerotherm, Barnes and Forberg improved the functionality and appearance of the Pan Am economy passenger chair and gave it a slimmer profile in the process:

> They specified, for instance, different densities of foam rubber in the headrest to minimize the lumpy contouring. They substituted a separate triangular foot bolster for the traditional bar footrest . . . And they were responsible for the hard-form plastic seat back with its attached tray. For the rest they simply added refinements – slight

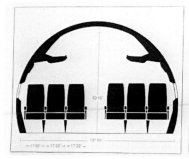

The modern airliner cabin took its form in the 1950s. Despite certain refinements, such as enclosed overhead storage lockers, things have changed very little. The elegant cabin of the BAC VC10 jetliner was outfitted by Robin Day in 1965.

changes in the seat buckle, in the ashtray, and in the armrest, and a new napkin for the headrest. Barnes's total design, in the last analysis, is patient attention to endless small details.[54]

However rational and well designed the seat and cabin of the early jetliners were, the experience of commercial flying after the 1960s became increasingly degraded in the eyes of many passengers. The simple fact that air travel had become mass travel made individual passengers feel unimportant: the novelty of flying was also an inevitable casualty of the mass travel market, and the standardization of service for economy passengers became subject matter for jaded stand-up comedians, such as the Australian stewardess-from-Hell, Pam Ann, and mock disaster films, such as *Airplane* (1980).[55]

To provide distraction, support and service, American airlines introduced female flight attendants in the early 1930s, at first employing only registered nurses in order to allay the fears of nervous passengers and to attend to those passengers who were made ill by the sensations of flying.

These young women were perceived to be so capable and well trained for their jobs as hostess, waitress, housekeeper, guardian of safety and angel of mercy that they also became objects of lust and attractive targets for parody.

With the development of the Boeing 747 wide-body jet, designed by a team including the production engineer Joe Sutter, Milt Heinemann and Frank Del Giudice and introduced in 1970, the increased size of the cabin enabled them to transform the space from a tube to a series of large rooms with nearly vertical walls and high, flat ceilings, from which the first enclosed overhead storage lockers were suspended. This was an entirely new kind of space with its own very particular characteristics, as unlike any previous aeroplane as it was unlike ships or trains. Ten abreast seating (three–four–three) in rows separated by two long aisles gave everyone in economy class a very clear impression of just how many people were on board and where each fitted into this gigantic puzzle. Many have experienced the full horror of a middle seat. Yet, with up to 568 seats, the 747 represented what Clive Irving called 'a quantum advance in airplane economics . . . expressed quite simply in one measure: the seat per mile cost, what it cost to carry a filled seat one mile . . .

In a mock-up of the Boeing 707 cabin (below), designers Walter Dorwin Teague and Danforth Cardozo discuss the layout and details that would become the industry standard for many decades. The great nave of the Boeing 747 jumbo jet (above) was an ideal shape to be used as a cinema in the years before miniaturization of video technology made viewing and listening a more private affair.

for the 747: a seat-mile cost 30 percent lower than the best achieved by the 707'.[56] The 747 started the cheap-flight revolution.

Yet there was still luxury to be had aboard the 'Jumbo', as the 747 came to be known. Up a sleek spiral stair was a world apart from the mechanistic repetition of the economy deck. Above the herd was a lounge for special customers: the famous and rich. The spiral stair was directly descended from earlier flying boats and the Boeing Strato-cruiser, which had featured just such an entrance to its swanky 'stateroom', a cocktail and smoking lounge below the main cabin. According to Irving, 'that stairway would, in fact, become part of the 747's signature, belonging in style to one of those Manhattan duplexes in a Sinatra movie'.[57] While Boeing planned the overall layout of the cabin and designed its essential features, consultant designers were hired by carriers such as Singapore Airlines or Air India to develop livery, interior colour schemes, upholstery, carpets, crew uniforms and details such as cutlery, all intended to give their cabins a distinctive flavour associated with the airline's place of origin. Like the great shipping lines of the past, national air carriers defined their visual images through national symbols and styles.

One of the most distinctive 747s and one closely identified with national image is the American presidential aircraft, Air Force One. The first Air Force One was a 707 commissioned by President Kennedy in 1962, with interiors and livery designed by Raymond Loewy Associates. The plane's appearance was conceived to represent Kennedy's New Frontier around the world. It was known as 'the flying White House', but it also served as the presidential hearse and the site of Lyndon Johnson's presidential inauguration after John F. Kennedy was assassinated on a visit to Texas in 1963. This echoed the debut of Pullman's first great railroad car, *Pioneer*, used as the hearse for the slain President Lincoln in 1865.

Interiors of the current presidential 747 appear as much like rooms in the White House as is possible within an airborne vehicle. The conference room features an enormous cherry-wood table, surrounded by large leather armchairs that swivel and rock. The public lounges, broad passageways furnished with sofas and armchairs, spacious stair hall and presidential suite are panelled with light wood veneers and appointed like typical corporate interiors in a modern high-rent office building. Yet the press cabin is furnished with three-abreast leather sofas, facing forward, aircraft-style. And the distinctive window-walls throughout the plane reveal unmistakably that this is a Boeing 747 Jumbo jet and not the large Washington mansion or Hilton hotel it resembles.

In the commercial realm, Frank Del Giudice produced some of the most extravagant 747 interior schemes in his early concept designs for

The Tiger Lounge proposal for the first-class section of the Boeing 747 stands out for its exoticism. Corporate bean counters were not impressed, and this scheme remained a prototype.

Boeing. One particular proposal, for the first-class lounge of the 747, stands out for its exoticism. This full-scale mock-up was known as the Tiger Lounge since it featured burnt orange carpets and continuous, wrap-around banquette seating upholstered in tropical foliage patterns and wild animal prints. It had a sunken bar area adjacent to a library/reading room, beyond which was a raked cinema. Airline bean counters dismissed all such generous uses of space, even for the first class. And the quaint idea that jumbo jets could be made into huge cinemas was demolished by the advent of the Walkman, personal computer games and the individual audio-video entertainment units and LCD display panels in seat backs and armrests. Ultimately, the in-flight entertainment programme supplanted the aerial view for all wide-body aeroplanes, in which only around 20 per cent of passengers were seated next to a window.

But before the sedative of the liquid crystal display arrived on the backs of airline seats, flight attendants presented a floorshow, themed around safety instructions and food service, their choreography set against the scenery of the cabin interior, which some airlines treated more theatrically than others. The best-known example of extreme aerial decor was the work of Braniff Airways, an American Midwestern airline, founded in the 1920s, which expanded to international routes in the 1960s. Braniff went against all conventional airline wisdom in 1965, when it introduced its new international image, conceived by the President, Harding L. Lawrence, and his wife, Mary Wells.

They employed the designer Alexander Girard, who headed the textiles division of the Herman Miller furniture company, to design their planes' vividly multi-coloured cabins and exterior colour schemes. The sculptor Alexander Calder created extravagant exterior graphics for two planes of the fleet. Most sensationally, the Italian couturier Emilio Pucci designed high-fashion outfits for the female flight attendants, who engaged in a notorious 'flight strip' by gradually removing layers of their uniform between take-off and landing. In bad weather the Braniff 'uniforms' were topped off with a clear plastic 'space bubble', replacing the traditional pillbox hat for rain protection in the years before the introduction of jet bridges, which linked the plane with the terminal. Their campaign was billed 'The End of the Plain Plane'. Unfortunately, lime green, plum, hot pink and burnt orange did not wear well in the heavily trafficked interiors of the jetliners, and the company had to replace them with more conventional and hard-wearing alternatives. Strongly styled interior designs have now gone, with the demise of independently minded airlines such as Braniff.

While highly regulated seating specifications for economy passengers have remained much the same for several decades, the average size of those who fly has not. The increasing girth of the typical passenger has meant that many people who fly are no longer able to fit comfortably into their seat. As early as 1969, the 747's design team realized that the standard seat width provided on the earlier 707 was not sufficient for the new, wide-body traveller, yet the economics of the new plane quashed negotiations to increase seat widths, which remain in 2005 the same as those of the 1950s, 17 inches between armrests. Some advances have been made, however, to achieve greater comfort within the old dimensions. The use of innovative design and new materials, including variable density foams, replacing traditional foam rubber, offer economy passengers better lumbar support in less bulky chairs, although the distance between upright seat backs (seat-pitch) has stayed at around 32 inches. The latest generation of economy airline seats offers winged headrests, adjustable neck supports, padded armrests, more refined seat-back storage pockets and adjustable footrests.

The most significant change, however, for those riding at the back of the plane, was the introduction of personal entertainment systems at every seat. The process began with the 747. According to Clive Irving, 'Given the space in the cabin, the temptation was to provide passengers with as many distractions as possible from the ennui of hours in the air and confinement to the serried rows. The airplane seat became a personal entertainment centre.'[58] Now, the airline passenger was cocooned in an electronic world of music, films, radio and television on demand, video games and telephone service. And many passengers use their own laptops, iPods or Gameboys. Once on board, a passenger can disconnect

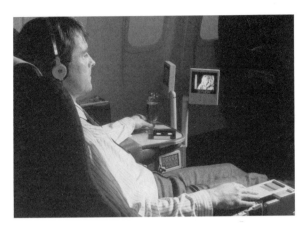

Personal entertainment, individually controlled from the passenger's arm-rest, is one of the main keys to comfort in jet travel today: the creation of a private world, insulated from the reality of riding for many hours in a metal tube tightly packed with other people.

almost completely from the community of the cabin and from the larger world outside the plane to focus, instead, on the deeply personal world of electronic fantasy or virtual reality. However, it is the very lack of physical activity made tolerable by electronic equipment that also puts passengers at risk of swollen feet or even deep-vein thrombosis.

These risks are not a problem for the wealthy minority who ride at the front. For British Airways intercontinental flights, for example, the company commissioned the consultancy Tangerine Design to create an individual pod compartment for each of their fourteen first-class passengers, 'designed as an ultra-luxurious efficiency centre . . . to provide living space, sleeping accommodation, dining, work and entertainment facilities' (see p. 10).[59] All the pods are angled towards the windows in the tapering nose of the plane, providing generous outside views and limiting direct visual contact with what is happening inside the cabin. Within the pod, tables, television, guest seat, storage compartments and power sockets surround a large armchair, which converts electrically into a flat 78-inch bed made up with fresh linen and duvet. The finishes are suitably smart, their decor involving cows and trees.

Recent business-class interiors in the major airlines have acquired the trappings previously associated with first class, such as paired, over-sized armchairs with flat or nearly flat reclining mechanisms, privacy screens, stand-up drinks and snack bar, and fully equipped work stations. Seats, still arranged in rows, are now 20 or more inches in width with 5-inch leather-covered armrests containing ever-increasing arrays of entertainment options and the controls for a wide range of seat adjustments: recline, headrest, lumbar, legrest, seat cushion and one-touch landing preparation.

Business-class upholstery is universally High Executive in style, with combinations of leather and cloth in dignified solid colours: blues, greys and black are favourites with the major airlines. Virgin Atlantic makes an exception by using tan leather, brushed metal and bright primary-coloured accents to project a more youthful and fashionable image. Typically, Qantas provides business-class seating for 80 passengers on their 747s. Since 2000 pod-type seats have been introduced in business class by British Airways, who may be followed by others, particularly in the cabins of the forthcoming generation of super-jumbo jetliners.

## Speed and Space

At the end of the twentieth century the only luxury commercial airliner still in operation was the Anglo-French Concorde, which flew for 25 years at supersonic speeds on the world's richest long-haul routes, mainly the North Atlantic. Although one might argue that its distinctive sound was intrusive near airports, and, from a democratic perspective, it was a symbol of indulgence and elitism, Concorde was such an extraordinary technical accomplishment and had such elegant lines that it was easy to accept. Its external grace was not, however, always matched by the slightly cramped and conservatively furnished passenger spaces inside its narrow, tubular fuselage. Strangely, Concorde was like the long tail of pre-jet flying. Its passengers were as economically privileged as those who flew before the Second World War, while the shape and size of the interiors harked back to the tightly tubular aircraft cabins of the 1930s.

Inside Concorde, passengers sat in paired seats flanking a central aisle. The circular section of this fuselage was as evident as it had been inside the DC-3. Small windows were set into wall panels similar to the PVC laminates used in other jet airliners of the period. Yet despite these familiar qualities, this was an aircraft of a different stripe. Technical issues associated with supersonic flight, such as added g loadings, demanded a fundamental redesign of the passenger seat. Concorde's seat could withstand tremendous stresses and was lighter by a third than the standard seat of the day, but in the British versions it retained an appearance of plush comfort and was originally upholstered in fabric reminiscent of London Underground trains of the 1930s. Given the many technical challenges, it is remarkable that the interior designers at Aérospatiale-British Aircraft Corporation were able to make the cabin of Concorde appear so ordinary.

Early Air France Concordes had more sophisticated interiors designed by the French offices of Raymond Loewy, CEI. These were sleek and fashionable cabins, their seat covers in fabric of various colours arranged randomly throughout the space, like jelly-beans, creating a light and playful atmosphere. Late in their careers, the interiors of British Airways Concordes were redesigned by Conran & Partners with the London consultancy Factory Design. The new cabins were brighter in colour and featured improved artificial lighting; toilets and galleys were upgraded using fashionable finishes, such as stainless steel; the seats were made lighter in weight through the use of titanium, aluminium and

Raymond Loewy's interior designs for the Air France Concorde fleet were significantly more sophisticated than those flown by British Airways. The decorative tray-table latches, crisply contoured surfaces of the seat backs and the geometry of the headrests are all typical of Loewy styling. He was particularly proud of the ceiling treatment, which created an illusion of greater width.

carbon fibre. Although the new chairs appeared sleek in the manner of an Audi 8, their dark blue Connolly leather upholstery identified them with the clubroom and the boardroom. Concorde's decor could never have matched its startlingly futuristic form, unless it had been designed like a set from the film *Barbarella* (1968).[60] Nevertheless, those who flew at twice the speed of sound aboard Concorde enjoyed an extraordinary view of the Earth's curvature from an altitude experienced only by an exclusive group made up of Concorde veterans, the world's elite fighter pilots and astronauts.

For those who have flown in space (Russians and Americans were the first), design has meant primarily the technical improvement of very crude and inhospitable vehicles. Early space capsules had to prove themselves capable of withstanding tremendous temperatures and compensating for enormous g-loadings and the effects of weightlessness on their pilots. The Soviets took the lead in space exploration with an astounding orbital flight in 1961, flown by the world's first cosmonaut, Yuri Gagarin, at the helm of *Vostok 1*. Yet Gagarin's flight and those of his Soviet successors were shrouded in secrecy. Neither the chief designer of *Vostok 1*, Sergei Korolev, nor anyone else involved in the programme was identified publicly at the time, nor was there much information made public about the vehicle itself.

Reports on the orbital flights made by the first generation of NASA astronauts were much more forthcoming, but their experiences sound surprisingly ordinary. According to Tom Wolfe, when John Glenn circumnavigated the globe in 1962 at 17,500 miles per hour in *Friendship 7*,

> He was sitting in a chair, upright, in a very tiny cramped quiet little
> cubicle 125 miles above the earth, a little metal closet, silent except
> for the humming of its electrical system, the inverters, the gyros, the
> cameras, the radio . . . He was supposed to radio back every sight . . .
> And yet it didn't look terribly different from what he had seen at
> 50,000 feet in fighter planes.[61]

Alan Shepard, who had been hurled into space on the previous mission, in a similar capsule atop a gigantic Mercury-Redstone rocket,
described the spartan, mechanistic interior of his vehicle in domestic
terms, as 'a busy little kitchen . . . whirring and buzzing and humming
along . . . there was nothing new going on!' From his capsule, Shepard
had viewed the earth only in 'a low grade black-and-white movie'
through his periscope, which contained a monochrome filter. But Glenn
had a window from which he could look directly at the earth below him:

> The clouds began breaking up over Australia . . . Off to one side he
> could make out the lights of an entire city, just as you could from
> 40,000 feet in an airplane, but the concentration of lights was
> terrific. It was an absolute mass of electric lights, and south of it
> there was another one, a smaller one . . . It was midnight in Perth
> and Rockingham, but practically every living soul in both places had
> stayed up to turn on every electric light they had for the American
> sailing over in the satellite.[62]

Thus, although the sight may have been familiar, the occasion was special. And through the tiny window of his capsule he also saw entirely
unexpected sights for which flight simulations had not prepared him:

> I am in a big mass of some very small particles that are brilliantly lit
> up like they're luminescent. I never saw anything like it . . . They're
> coming by the capsule, and they look like little stars. A whole shower
> of them coming by. They swirl around the capsule and go in front of
> the window . . . I can see them all down below me also.[63]

The early astronauts had to pay constant attention to their control
panels, which covered much of the available surface inside the ship. The
rest of the interior was a maze of wires and cables against a background
of military grey paint. Views out were of little importance to the success
of a flight, except for their considerable public relations value, which

Glenn understood well. When NASA assigned astronauts to contribute their experience to the design of new spacecraft, he took special responsibility for the arrangement of controls and the general layout of the cockpit. This collaborative approach to the development of space-capsule interiors was extended when NASA retained Loewy and his partner, William Snaith, as interior design consultants for its Skylab orbital space station project. By doing so they demonstrated a commitment to designing ships' interiors that might improve the quality of life for the astronauts, who would be spending extended periods of time in space. Yet the window remained a point of contention in spaceship design throughout the 1960s, since it created constructional complications and therefore added cost.

Skylab was the first American space station. Launched in 1973, it spent 171 days circling the earth and provided accommodation for three crew members, the longest serving of whom remained aboard for 84 days. The station was launched into orbit by a Saturn V Moon Rocket, and provided accommodation, laboratory and workshop facilities in two adjoining compartments. Although it had the same volume as a small house, it was packed with scientific equipment for the solar and microgravity experiments performed by the crew as they circled the earth. For this reason, it retained the mechanistic appearance of earlier orbital capsules, with most surfaces covered in instrumentation and control panels, tubing, wiring and bare structural elements. The significant difference from the earlier orbital capsules was that it provided sufficient space for the astronauts to move around in the performance of their duties, for the rituals of personal hygiene, to exercise and for other domestic activities.

As a result of his extensive 'habitability studies', Loewy contributed a number of features to the design that made the station a more comfortable and congenial living environment for the crew. One of these was the provision of a round window. Loewy wrote:

> the inclusion of a porthole was essential. I stated that I could not fully endorse a capsule in which there would be no possibility for months at a time to look out and see our earth. All the Skylab crews stated at the time of debriefing that without the porthole the mission might have been aborted.[64]

While the last line is typically self-congratulatory, Loewy's work yielded more than one improvement to the interiors of Skylab. They included more sophisticated storage furniture and adjustable seating arrangements in the crew compartment, which enabled the crew to dine facing one another. Their sleeping provisions were also made more pri-

'Up' and 'Down' are formless notions in NASA's Skylab orbital vehicles. The concept of ergonomics had to be seriously reassessed in the design of these vehicles.

vate and comfortable thanks to the Loewy office's interventions. Light-coloured and easily cleanable surfaces made 'housework' easier and more effective, especially in the event of uncontrollable space sickness, while a more orderly disposition of equipment facilitated circulation through the cramped interiors.

In the final tally, Loewy's work for NASA was of little consequence in comparison with the vast technological achievement of the project, but it was enough to establish the value of design for the quality of life inside space vehicles. Consequently, their ideas were applied more extensively in the design of subsequent space stations and, beginning with the *Enterprise* of 1976, the more aircraft-like shuttles that tended them.

## Today and Tomorrow

Returning to the field of commercial aviation, today's pundits predict that in the future airliners will be bigger, faster and more environmentally friendly. But it remains to be seen if they will be any more comfortable or enjoyable for passengers. Plans for the latest generation of large commercial aircraft are full of promises, but in the end the reality of economics may tell a different story. What is known at the time of writing is that the Airbus A380, a German-Spanish-British-French co-production, is the biggest plane in the world. It is also the first double-decker airliner since the Boeing Stratocruiser and the Breguet 763

Provence of the 1950s. It seats upwards of 550 passengers in three classes and is touted to address the problem of overcrowding in the skies (see p. 17).

Chris Voysey, who managed the project, brought together experts in many fields of aeronautics to produce an aircraft using the most advanced production technology and an innovative wing design, which was needed if such a large aeroplane was to be viable. One of the expressed aims of the project was to revitalize the passenger experience by providing a more spacious cabin in all classes, with room to move about during long flights. This may run into conflict with safety practices that encourage passengers to remain seated unless absolutely necessary. However, the plane, in its prototype configuration, includes double-width entry doors, a spacious reception area and a double staircase to the upper deck. Only the most optimistic believe that airlines will use the larger spaces of the A380 for generous passenger amenities, such as children's play areas, a gym, bar or lounges.

Hot on the heels of Airbus, Boeing has unveiled a plan for their answer to the A380. Called the 787 Dreamliner, the proposed super-jumbo will offer fourteen Pullman-style sleeper cabins and will accommodate 550 passengers. According to Boeing, this new plane will provide a 'passenger-focused interior . . . that will connect passengers as never before to the flying experience'.[65] According to the editor of *Aircraft Interiors International*, their designers will concentrate on the detail of larger long-haul aircraft interiors, providing sleeping cabins for business travellers, play areas for children, larger windows on the 787, and mood lighting in all new planes to mimic natural light at different times of day.[66] The hyperbole will have to be matched by an equally ambitious seat cost per mile for the airlines, if the arched entrance hall, atrium, simulated sky-lighting, wider seats, large windows and relaxing ambience are to awaken the pleasure of commercial flying for those in economy, as well as passengers in business and first class.

Looking further ahead, Boeing has projected a more radical design for an airliner derived from the B-2 Stealth Bomber. The 'blended wing-body aircraft' will seat 800 passengers, only 50 more than the number intended to fly in Howard Hughes's 'Hercules' flying boat, the *Spruce Goose*, which flew only once in 1947. Based on the same Northrop 'flying wing' design as the B-2, the new plane would have no traditional fuselage, but accommodate its passengers in a wing, two decks thick.

The chairs, pods and other furnishings of such aircraft may resemble in some ways the luxurious first-class sleeper beds by furniture designers such as Ross Lovegrove and Marc Newson. Newson's aircraft designs range from the interior of a small private jet, in his characteristically sleek futurist style, to the Qantas Skybed (with BE Aerospace), a smooth plastic shell-form reminiscent of Eero Aarnio's Globe chair of 1960. The Qantas chair converts into a bed 6.5 feet long and nearly two feet wide. It also features innovative luxuries, including a back massager, the first in-flight Short Message System (SMS), free telephone calls to other passengers and 10-inch TV screens for the personal entertainment system.

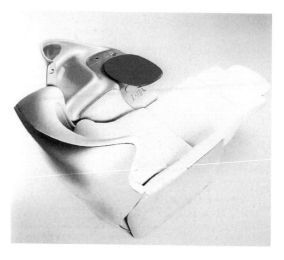

For Japan Airlines, the British designer Ross Lovegrove created the Skysleeper Solo seat, a curvaceous plastic shell structure that converts to a flat bed, with electrically adjustable back and footrest, massage function and pneumatically adjustable cushions. The left armrest erupts into an erotic white sculptural form incorporating a storage basin, a movable red tray-table and other features in bright green and yellow. Lovegrove described the organic design of the seat: 'It is sensual, curvaceous, and body hugging, without straight lines or sharp angles – just like the human body'.

These and other prototypes, experiments or production-model airline sleeper seats employ the most advanced materials and electronic technologies to improve the experience of flying for first- or business-class passengers. Some of their innovations will inevitably filter down to the economy seats in terms of construction, finish, connectivity and even elements of style, but not for space. Compensation will be provided by technology. In the future few passengers on board the blended-wing aircraft, for example, will have a window view, but they will see outside via a video display unit, which will also serve as their entertainment and communication facility. In terms of its general layout, this project is reminiscent of Norman Bel Geddes's Airliner No. 4, but its pleasures will be generated electronically rather than through the direct experiences offered by a promenade deck, ballroom and swimming pool. No one will dress for dinner.

Sleek and chic: the Skysleeper Solo business-class chair designed for Japan Airlines by Ross Lovegrove in 2002 adapts the visual characteristics of modern organic furniture design to the aircraft cabin.

# CONCLUSION

Mobility in the modern world can be classified as belonging to two types: transportation and travel. The former is concerned with getting from A to B in the most efficient and economical way possible, a trip often experienced as a void in time, a gap between places. The latter is about the quality of experience between leaving A and arriving at B. As vehicles and vessels have become more specialized over time, the relation between transport and travel has become increasingly complex in design terms. Initially, as in the case of nineteenth-century ocean liners, both could be achieved in the same vessel by the segregation of classes. Steerage passengers were transported, while first-class passengers travelled. Yet this distinction is sometimes blurred. The example of London Transport between the two world wars demonstrates how enlightened design can inject the pleasure of travel into the most common transporters used by the great majority of an urban population.

The big question for the twenty-first century is if or how the developed world can sustain the extent of mobility we now expect. In a world threatened by the looming effects of global warming, where suicide bombings make even urban bus travel a daunting prospect, where terrorists fly jetliners into skyscrapers, where indigenous cultures are jeopardized by mass tourism, and where work is increasingly conducted electronically, the virtues and advantages of mechanized travel are becoming increasingly questionable. Today, however, more people are moving around the world than ever before, and the trend to increasing mobility shows no evidence of slowing down without some serious impediment affecting it. The expectation, held by increasing numbers of people around the world, seems to be that holiday travel is a birthright. For business people it remains an economic imperative. For the military it is duty. And to political refugees, transport is survival.

Design continues to play an essential role in our expectations of modern mobility, whether at the mundane level of the school run or, like Hugh Hefner, in the fur-trimmed bedroom aboard his private jet. Established designers and manufacturers put their identifiable stamp of style on vehicles and vessels, lending them a visually recognizable guarantee of quality, such as Christian Lacroix's train interiors for the SNCF. Corporate identity also remains a significant factor in transport design, whether the most prominent identity is that of the designer, the manufacturer or the carrier, as in the cases of airlines and shipping companies – Virgin aircraft and trains project a youthful image, while designs for Singapore Airlines remain distinctively cultural. Thus, notions of brand loyalty continue to apply in transportation as they do in relation to breakfast cereals and clothing.

Alongside branding, national identity inflects the designs of many vehicles in which we travel. Although most cars today are the products of global industries, the latest automobile interiors still convey messages of national taste and pride. These symbols operate in a more condensed space than in the nationally branded ocean liners of the past – *Ile de France, Conte di Savoia, Mikhail Lermontov* – but the signals are comparable and just as effective. And they remain consistent over time. The dashboard of a 2005 Bentley is equally replete with symbols of Britishness as that of a 1955 model, woodiness and craftsmanship the hallmarks of these monuments to excellent engineering and economic privilege. Bentleys remain 100 per cent British in image, even though the company is now a subsidiary of Volkswagen AG.

The perpetual quest for ultimate physical comfort persists in all areas of transport design. From Pullman's convertible railway berth in the 1860s to Tangerine Design's iconic Club World Sleeper Seat for British Airways, the newest materials and technologies have been employed to provide the maximum ergonomic control over the travel environment, with the aim of achieving the greatest comfort attainable in the particular circumstances. Travellers have thus come to expect increasingly high standards of refinement and, in particular, adaptability to personal dimensions and postures, as in the intelligent drivers' seats in today's luxury cars or the highly adjustable air suspension seats in the cabs of long-distance lorries.

Psychological comfort and pleasure are also enduring aims of transport designers. A German study, conducted in the 1990s, revealed that 72 per cent of their respondents cited 'relaxation' as the main motive for their travels, while 67 per cent said that they travel to escape routine.[1]

With such attitudes in mind, designers of transport interiors may be right to emphasize the luxurious and exotic elements of their schemes. According to Marc Newson, whose stylish Qantas Skybed evokes the essential modernism of flight, 'The world of civil aviation appears to us as pragmatic and rational and boring, which is absurd when you think how exciting the experience of flying is.'[2]

And so, diversion and entertainment continue to motivate designers of transport interiors. On today's giant cruise ships extravagant decor and service come as standard, and a direct visual contact with the natural world outside is more carefully developed than ever before. By contrast, in high-density transporters, constrained by limited space and tight cost factors, digital media provide passable alternatives. The view from the window of an airliner is now less important than the view of the VDU screen, which provides images of glamour and escape independent of the actual flight. The personalized audio-visual facilities in economy class create a much-needed virtual environment for passengers on long-haul flights. Indeed, many jaded passengers will not bother to look up from the screens of their personal laptop computers or games consoles to take in even the most spectacular views of earth and sky.

Changing tastes in sports and leisure activities are also reflected in transport design. Rock-climbing walls have become a prominent feature on the decks of modern cruise ships, somewhat displacing traditional activities such as shuffleboard. Meanwhile, disco nightclubs are now the focus of technically elaborate designs and extravagant lighting effects, overtaking ballrooms in popularity among a growing clientele of younger cruise passengers.

Historically, comfort and delight have frequently been compromised by the demand for utility, as illustrated by the demise of the London Routemaster bus, its replacement a characterless Eurobox capable of transporting considerably greater numbers of standing passengers and wheelchair users, but delivering little comfort or opportunity for pleasure. In the most extreme emergency circumstances during periods of war, the transformation of luxurious ocean liners into vastly overpopulated troop carriers involved the application of strict utilitarian principles to interior planning. Travel today for most people in less developed parts of the world remains utilitarian, if not harsh, giving the Western term 'basic transportation' an altogether greater significance than in richer parts of the world.

Some of the most singularly important vehicles in history have also been the most utilitarian in design. Lindbergh's *Spirit of St Louis* had as austere a cockpit as any plane ever built, but it became an icon of progress in the twentieth century and the compelling site of the Lone Eagle's widely read memoir of the first non-stop transatlantic flight. Speedboats, racing cars and spacecraft are also designed with technical parameters overshadowing comfort or appearance as design priorities, although sufficiently refined ergonomic factors in the design of seats and controls remain important to the driver's or pilot's performance.

The accounts of early air travellers suggest that a certain level of discomfort was part of the pioneering experience, since some passengers chose to fly in open cockpits, for the thrill of the wind in their faces and the roar of the motor, over the dubious comforts of enclosed cabins. Similarly, many of today's sophisticated one-design class sailboats offer crew members little creature comfort in favour of an experience that is robustly energetic and at one with the elements of wind and water.

While thrills are still possible in modern transport vehicles, design for safety has become a top priority of manufacturers, regulators and occupants of anything that moves. From the disgraceful shortage of lifeboats on the *Titanic* to the timid introduction of seat belts in some automobiles of the 1950s, passenger safety was a hard-won concept against popular indifference and the priority of corporate profits in a world of laissez-faire business. Ralph Nader and other crusading idealists finally forced the issue in the American motor industry, but industry regulation and formal design education were ultimately the most effective means of changing the culture of transport design and establishing a serious regard for designed-in safety. Coventry University, the Art Center College of Design in Pasadena and the Domus Academy in Milan are among many institutions offering undergraduate and postgraduate courses in Transportation Design, which now set challenging curricula for student designers.

Regulations too play a huge part in advancing travellers' safety. IATA was one early initiative. Another is the Australian Design Rules, developed through a partnership between government, industry and consumer bodies, which established nationwide automobile seat-belt laws by 1973, quickly reducing front-seat fatalities by 45 per cent. They continue to set national standards for car passenger and pedestrian safety and for $CO_2$ emissions.

Perhaps the most elaborate safety technology is found today in travel chairs, such as the McDonnell Douglas rocket-catapult ejection seat, designed for jet fighter pilots. Ejection seats, pioneered by Saab in Sweden during the early 1940s, influenced the design of passenger car seats, child safety seats and even the fanciful gadgetry of James Bond's Aston Martin DB5, from which Sean Connery famously ejected a tiresome villain in the film *Goldfinger* (1964).

The interiors of automobiles are now far more defensive places than they were even for Connery's Bond. The tiny, French-built Smart car is constructed around a 'tridion safety cell', a strong steel cage designed to work in conjunction with the seats, seat belts, airbags and even the tyres to protect its occupants. Modern ships also are far less prone to disaster than in earlier times thanks to satellite navigation and echo sounding, as well as new interior fireproofing techniques. Yet even the best design cannot defend passengers against mismanagement or outright recklessness, whether the perpetrator is a major international shipping company or a drunk at the helm of a speedboat.

At a moment in history when vehicle interiors have become ergonomically sophisticated spaces in which many people spend large amounts of their time, we face an equally vast crisis in the viability of our current transportation needs and desires. We have come nearly to a point where political agendas regarding the control of oil supplies, national and global economics, the control of the transport infrastructure (highways, rail track, air lanes and landing runways) and popular expectations of personal mobility may collide with the realities of climate change and diminishing natural resources. The balance between our travel desires and transport needs may have to shift, and this could significantly affect our experiences of travelling and of moving from A to B. Cheap flights over short distances are currently diverting passengers from more efficient and less polluting high-speed train routes, while the global craze for big, defensive automobiles has overwhelmed the development of small, cleaner, more fuel-efficient private cars. All this results from a cocktail of commercial pressures and the preference of individuals for privacy, independence, economy and image.

Breakthrough technology, such as the innovative propulsion systems of hybrid, petrol-electric automobiles, has already had a subtle impact on vehicle interiors. The dashboard display of the Toyota Prius hybrid, for example, concentrates the driver's attention on the car's economical delivery of power, through a touch-screen graphical display, the energy monitor, which dominates the dashboard and also controls satellite navigation and the audio entertainment system. More dramatically, the fully digital cockpit of the AH-640 Apache Longbow attack helicopter, which has been in service since 1998, links with a display in the pilot's helmet; wherever the pilot looks, the guns look, making war seem like a giant video game.

Our increasingly deep interaction with digital entertainment and communication equipment, the cell phone and the MP3 player, help to make high-density transport vehicles bearable environments, especially over very long distances, but also for short urban hops. Yet they are also indispensable in the most demanding travel experiences, such as sailing around the world single-handed. Dame Ellen MacArthur made advantageous use of the electronic equipment on board her 75-foot carbon fibre trimaran, B&Q Castorama, during her record-breaking voyage of 2004–5. Her activities included frequent conference calls, via satellite telephone, with her global team of tactical advisors, access to weather websites and a personal 'weather router', and consultations with on-call medical, nutritional and sleep advisors. In the event of a problem, she had an array of electronically deployable safety devices, and, for a bit of rest, an autopilot.

As noted above, one of the great promises of modern vehicles and vessels is the pleasure they offer us. The next generation of airliners provides a useful signpost to what we may expect. Airbus and Boeing both claim to have the passenger experience at heart in their next generation of planes. However, it is undoubtedly the cost-per-seat-per-mile of running these leviathans, combined with technological advances affecting passenger comfort, that will determine their eventual interior layout and amenity. Pleasure has a price.

In the early twenty-first century two of the greatest threats to the ecology of the planet are the automobile and the jet aircraft. In 2006 transportation vehicles produced half of all British carbon emissions. Aviation, alone, is the fastest growing single contributor to climate change. The cost of one round-trip flight (one seat) between Europe and the USA produces more carbon dioxide than the average automobile emits in one year. Yet, how else do you get there? Perhaps there is a case for long-haul flying even at current subsidized rates. Short-haul flying, on the other hand, is much harder to defend when the alternative train trip is substantially less damaging to the environment.

According to Tony Collins, Chief Executive of Virgin Trains, modern intercity locomotives produce one-tenth the carbon dioxide emissions of short-haul aircraft. The Eurostar from London to Paris emits one-eighth the carbon dioxide as the same trip by plane. City centre to city centre, the train is faster, and it is potentially more comfortable, although more expensive because of zero-rated tax on aviation fuel. The US Department of Energy concurs. It reported that in 2003 Amtrak was 18 per cent more energy-efficient than commercial airlines. Its energy intensity was 2,935 BTUs (British Thermal Units) per passenger mile as compared with 3,587 for the airlines and 3,549 for automobiles. A further benefit is that Amtrak uses diesel fuel produced at a higher volume per barrel of crude oil than commercial aircraft, which burn more highly refined fuel.

A growing ecological consciousness among designers and manufacturers has yielded some modest results. The strong weight-saving composite sheet materials and variable density foams, used increasingly in aircraft interiors, will have a positive effect on fuel efficiency. Biodegradable plastics, used in car interiors, will help to limit the mountains of waste material building up all over the world. A return to natural materials, including timber, bamboo and woven natural fibres, is displayed in the interiors of the latest Japanese Shinkansen trains, if only to symbolize the ecological advantages of the express train over the aeroplane for journeys up to 400 miles. In a similar spirit, a 2004 award-wining British transportation design student applied IKEA's environmentally friendly design and manufacturing policies to a concept car interior: 'IKEA's mission is to bring exclusivity of design to the majority and I tried to incorporate this into the interior of the car. I think ecological design could be the way forward for car interiors in the future', said Reg Shola Hingston. Dried banana leaves and rattan on a metal frame formed the car's dashboard, vents and sides of the centre console. Door panels and console top were finished with birch, a common sustainable timber, and leather upholstery provided luxurious comfort while avoiding the use of synthetic fabrics.[3]

The style of our mobile living rooms, whether capsules, pods, tubes or boxes, continues to blend modernism and tradition in new ways. Security and familiarity remain important constituents in determining the way transport vehicles look and feel inside. Something of the luxury of the earliest royal trains, carriages, yachts and ocean liners persists in the wood-panelled and sumptuously upholstered cells in which we go about. Yet the effects of technological progress are ever present, not only in the equipment we use to control our vehicles and entertain ourselves while on the move, but in our changing taste for the sleek and new, for shapes and materials that express the ingenuity of the machine itself.

Our current attraction to retrofuturist styling does not simplify matters. Today, some of us may very well find ourselves flying in facsimiles of the camp, fur-lined spaceship in which the super-heroine Barbarella brought erotic love to the galaxy in Roger Vadim's comic sci-fi epic of 1968. In trendy bars and upmarket travel agencies, we commonly enjoy cool, Italian-inspired futurist interiors reminiscent of the Pan Am space shuttle featured in Stanley Kubrik's film *2001*.

Yet our belief in unlimited progress is thrown into sharp relief by the uncertainties of today's troubled world. It is legitimate to ask how long the human race can pursue unimpeded travel at the current rate of impact on natural resources. Design innovation can help, if the brief is good, and it may yet happen that comfortable and convivial magnetic levitation (MAGLEV) trains replace most regional inter-city air traffic. They may link sparklingly clean futuristic cities entirely covered by transparent geodesic domes that ensure a perfect climate for their privileged inhabitants, who could ride over cycle-superhighways on sleek, carbon fibre, bubble-topped bicycles with full connectivity and sat-nav, the last word in personal transportation.

# REFERENCES

## 1 LAND

1  Jacques Damase, *Carriages, Pleasures and Treasures* (London, 1968), p. 77.
2  Ibid., p. 58.
3  J. R. Kist, *Daumier: Eyewitness of an Epoch* (London, 1976), p. 85.
4  Charles Baudelaire, *The Painter of Modern Life and Other Essays* (New York, 1964), pp. 38–40.
5  Ibid.
6  Ibid., p. 81. John Loudon McAdam perfected a system of road surfacing first developed in the late eighteenth century by Trésageur, a French Inspector of Highways under Louis XVI. Macadam roads had a foundation of large stones laid with a camber or slight convex curvature to ensure that rainwater would drain off. The foundation was then covered with crushed stones bound with gravel for a smooth surface.
7  *www.victorianlondon.org* [accessed 20 March 2005].
8  T. C. Barker and M. Robbins, *A History of London Transport* (London, 1963), vol. I, pp. 84–5.
9  Ibid., p. 230. Fares fell from an average of around 4d in 1860 to an average of around 1¹/₂d in 1889.
10  Ibid., p. 295.
11  'The Trolley Song', *Meet Me in St Louis*, Hugh Martin and Ralph Blane (USA, 1944).
12  Booth Tarkington, *The Magnificent Ambersons* (Garden City, NY, 1918); *www.bartleby.com/160/*, chapter 1, para. 12 [accessed 20 August 2005].
13  This is the company's own claim: *www.studebakermuseum.org* [accessed 13 September 2004].
14  For a thorough discussion of the American system of production, see David A. Hounshell, *From the American System to Mass Production, 1800–1932* (Baltimore, MD, and London, 1984).
15  'The Surrey with the Fringe on Top', *Oklahoma*, Richard Rodgers and Oscar Hammerstein (USA, 1943).
16  Tarkington, *The Magnificent Ambersons*, chapter 1, para. 2 [*www.bartleby.com*, accessed 20 October 2004].
17  *Illustrated Catalogue of the Great Exhibition London, 1851* (reprinted London, 1970), p. 110.
18  Siegfried Giedion, *Mechanization Takes Command* (Oxford, 1975), pp. 457–8.
19  Corin Hughes-Stanton, *Transport Design* (London and New York, [1967]), p. 8.
20  'Train Impérial', *L'Illustration* (1857); reprinted in Giedion, *Mechanization Takes Command*, p. 456.
21  Ibid., p. 448.
22  Elizabeth Collins Cromley, 'Sleeping Around: A History of American Beds and

Bedrooms', The Second Banham Memorial Lecture, *Journal of Design History*, III/1 (1990), pp. 1–18.

23  See Wolfgang Schivelbusch, *The Railway Journey: The Industrialization of Time and Space in the Nineteenth Century* (Los Angeles, 1986).

24  Giedion, *Mechanization Takes Command*, p. 452.

25  Ibid.

26  Russell Lynes, *The Tastemakers* (New York, 1954), p. 96.

27  Dr J. Héricourt, *L'Hygiène moderne* (Paris, 1907); quoted in Adrian Forty, *Objects of Desire* (London, 1986), p. 163.

28  'Star Track: The Return of the Orient Express', British *Vogue* (London, December 1981), pp. 203–7.

29  Paul Cret, 'Streamlined Trains', *Magazine of Art* (1937), pp. 17ff; reprinted in Claude Lichtenstein and Franz Engler, *The Esthetics of Minimized Drag, Streamlined: A Metaphor for Progress* (Zurich, n. d.), pp. 268–9.

30  Ibid.

31  G. Votolato, 'Raymond Loewy', *The Interior* (Melbourne), I/2 (September–November 1991), pp. 14–15.

32  Raymond Loewy, *Industrial Design* (London and Boston, MA, 1979), p. 87.

33  '50 Years between Trains', *Popular Science Magazine* (June 1952), pp. 100–01.

34  Lichtenstein and Engler, *Esthetics of Minimized Drag*, pp. 52–3.

35  Ibid., p. 130. Stout's Railplane was a considerably more practical prototype than Kruckenburg's design, since its diesel engine was linked by automatic transmission to the driving wheels, creating a safe and simple propulsion system.

36  Martha Thorne, *Modern Trains and Splendid Stations: Architecture, Design and Rail Travel for the Twenty-first Century* (Chicago, 2001), pp. 26–31.

37  Sarah Stanczyk, ed., Translink Press Release (January 2003): *www.translink.co.uk*.

38  Thorne, *Modern Trains and Splendid Stations*, p. 54.

39  For an introduction to the commercial aspects of London Transport's design policy, see 'Design and Corporate Identity', in Forty, *Objects of Desire*, pp. 222–38.

40  Ibid., pp. 224–5.

41  Christian Barman quoted by Nikolaus Pevsner, 'Patient Progress: The Life Work of Frank Pick', *Architectural Review*, XCII (August 1942), p. 31.

42  Enid Marx, quoted by Oliver Green and Jeremy Rewse-Davies, *Designed for London: 150 Years of Transport Design* (London, 1995), pp. 62–3.

43  *Midnight Cowboy*, dir. John Schlesinger (USA, 1969), based on the novel by James Leo Herlihy.

44  This passage on Kesey was first published in G. Votolato, *American Design in the Twentieth Century* (Manchester, 1998), p. 62.

45  'He Drives His House to the Country', *Popular Science Magazine* (June 1952), pp. 88–90.

46  Karal Ann Marling, *As Seen on TV: The Visual Culture of Everyday Life in the 1950s* (Cambridge, MA, 1994), pp. 3–5.

47  Paul Virilio, interviewed on 21 October 1994 by Louise Wilson, for CTheory: *www.ctheory.net/text_file?pick=62* [accessed 4 October 2004].

48  Ilya Ehrenberg, *The Life of the Automobile* (London, 1985), p. 11.

49  Herkomer, quoted in Gerald Silk, *Automobile and Culture* (New York, 1984), p. 75.

50  Judith Hoos Fox, ed., '2wice', *Inside Cars*, V/2 (Princeton, NJ, 2001), p. 62.

51  Gaston Rageot, quoted in Paul Virilio, *The Aesthetics of Disappearance* (New York, 1991), p. 61.

52  Roland Barthes, *Mythologies* (London, 1973), pp. 96–7.

53  James Wolcott, in Judith Hoos Fox, '2wice', p. 53.

54  These photographs are collected in Mell Kilpatrick, ed. Jennifer Dumas, *Car Crashes and Other Sad Stories* (Cologne, n. d.).

55  Votolato, *American Design*, pp. 120–24.

56  This is now a layout used in supercars, such as the McLaren F1 high-performance sports coupé: *www.mclarencars.com*.

57  For a detailed cinematic tour of the interior of the 1958 Thunderbird, see *Mean Streets*, dir. Martin Scorsese (USA, 1973).

58  Wert Bryan, 'Dynamics of Car Seat Design', *Illumin*, v/1 (Fall 2005).

59  Teague quoted in Stephen Bayley, *Sex, Drink and Fast Cars* (London and Boston, MA, 1986), p. 78.

60  John Gartman, 'Wired News: Women Drive Changes in Car Design': *www.wired.com/news/autotech/0,2554,62991,00.html* [accessed 31 October 2004].

61  Martin Buckley, 'Rococo Roller', *Daily Telegraph* (16 October 2004), p. P2.

62  This passage on the 1958 Cadillac was first published in Votolato, *American Design*, pp. 116–17.

63  'Retrofuturism' is a term used to describe the millennial tendency in car design to the revival of earlier models, such as Mini and Beetle, and the fusions of traditional forms and details with advanced technology. See Brook Hodge, *Retrofuturism: The Car Design of J. Mays* (Los Angeles, 2002).

64  John Steinbeck, *Cannery Row* (London, 1958), p. 135.

## 2  WATER

1  All quotes from Charles Dickens, *American Notes* (London, 1842), chapters 1 and 2 [*www.bibliomania.com*].

2  W. S. Tryon, *My Native Land: Life in America, 1790–1870* (Chicago, 1961), p. 80.

3  Ibid., p. 182

4  N. Hawthorne, in 'Sketches from Memory', from *Mosses from an Old Manse* (New York, 1846); quoted in Siegfried Giedion, *Mechanization Takes Command* (Oxford, 1975), p. 459.

5  Tryon, *My Native Land*, p. 80.

6  Ibid., pp. 180–81.

7  Ibid., pp. 307–8.

8  Ibid., p. 182.

9  Russell Lynes, *The Tastemakers* (New York, 1954), p. 89.

10  The Blue Riband award was established in 1839 for the fastest Atlantic crossing of a passenger ship. It was last awarded to an ocean liner, the ss *United States*, in 1952, although the catamaran *Hoverspeed Great Britain* was the first of a series of high-speed ferry boats that beat the previous record in the 1990s: *www.blueriband.com/*.

11  John Heskett, *Design in Germany, 1870–1918* (London, 1986), pp. 99–105.

12  Jack Fritscher, *Titanic* (San Francisco, 1999), pp. 18–19.

13  The Olympic Class comprised three White Star Line vessels, the *Olympic*, *Titanic* and *Britannic*, all built by Harland & Wolff of Belfast. Tom McCluskie, *Anatomy of the Titanic* (London, 1998), p. 9.

14  Ibid., p. 148.

15  Ibid., pp. 127–8.

16  Le Corbusier, *Towards a New Architecture* (London, 1982), pp. 90–94.

17  Raymond Loewy, *Industrial Design* (London and Boston, MA, 1971), p. 104.

18  Edward Lucie-Smith, *Furniture: A Concise History* (London, 1979), p. 173.

19  Evelyn Waugh, *Brideshead Revisited* (New York, 1960), p. 217.
20  Corin Hughes-Stanton, *Transport Design* ((London and New York, [1967]), p. 27.
21  James Steele, *Queen Mary* (London, 1995), p. 96.
22  Ibid., p. 178. Capable of more than 31 knots, the *Queen Mary* was faster than any contemporary warships and, at least in the early years of the war, faster than enemy torpedoes.
23  Loewy, *Industrial Design*, pp. 163, 202.
24  Votolato, *American Design*, p. 141.
25  Leslie Reade, quoted in Steele, *Queen Mary*, pp. 186–7.
26  Gordon R. Ghareeb, 'A Woman's Touch: The Seagoing Interiors of Dorothy Marckwald': *http://home.pacbell.net/steamer/marckwald.html* [accessed 10 April 2005]
27  Transatlantic passage on the *Aquitania* and the *Mauretania* in 1925 cost $255 in first class, $150 in second class and $90 in third class. Cunard advertisement, *National Geographic Magazine*, xlvii/3 (March 1925).
28  The *Mikhail Lermontov* eventually retired from the Leningrad–London–Le Havre–New York shuttle service to become a cruise ship, operating mainly in the Antipodes, where the ship has lain since its sinking off the New Zealand coast in 1986. There, the intact liner has become a celebrated wreck and a diver's paradise: *www.nzmaritime.co.nz/lermontov.*
29  Colin Boyd, 'Case Studies in Corporate Social Policy', in James Post, William C. Frederick, Anne Lawrence & James Weber, *Business and Society: Corporate Strategy, Public Policy, Ethics*, 8th edn (New York, 1996).
30  Bruce Peter, 'Tag Wandberg and the Modern Cruise Liner', paper presented to the conference *Modern Voyages: Sea Travel Since Brunel: Bristol University, 20 April 2006.*
31  Le Corbusier, *The Decorative Art of Today* (London, 1987), pp. 68–91.
32  Cleveland Amory, *The Last Resorts* (New York, 1952), p. 290.
33  Carlos Baker, *Ernest Hemingway: A Life Story* (New York, 1970), pp. 332–3.
34  The us dominates popular boating, with one boat per 17 persons as compared with one boat per 61 persons in France. Statistics provided by us & Foreign Commercial Service and us Department of State, 2002: *www.exporthotline.com* [accessed 4 October 2005].
35  Malcolm Oliver, Travelpage.com: *http://www.cruiseserver.net/travelpage/ships/rc_brilliance.asp* [accessed 28 November 2004].
36  E. B. White, quoted in Alexis Gregory, *The Golden Age of Travel, 1880–1939* (London, 1998), p. 176.
37  Vittorio Garroni Carbonara, 'Futuristic Cruise Liners', paper presented at the *Seatrade Asia Pacific Cruise Convention: Singapore, 4–7 December 1996: http://www.cybercruises.com/garronispeech.htm* [accessed 8 October 2005].
38  Dr Tim Unwin, 'The Great Eastern: A Floating City', paper presented to the *Modern Voyages* conference.

## 3 AIR

1  Orville Wright and Wilbur Wright, 'The Wright Brothers Aeroplane', *Century Magazine* (September 1908): *http://www.wam.umd.edu/~stwright/family/presidents.html#contact* [accessed 15 January 2005].
2  Le Corbusier used this phrase to describe the spirit and character of the modern

engineer and architect: *Towards a New Architecture* (London, 1982), p. 119.

3 Tom Wolfe, *The Right Stuff* (London, 1981), p. 35.

4 Quoted in Robert Wohl, *A Passion for Wings: Aviation and the Western Imagination, 1908–1918* (New Haven, CT, 1994), p. 35.

5 In 1909–10 altitude records advanced at an astounding rate, from the first reliable record of 450 feet, achieved by Louis Paulhan flying a Voisin biplane in January 1909, to Armstrong Drexel's world altitude record of 6,745 feet, set in a Blériot in July 1910. Charles E. Vivian, *A History of Aeronautics* (Seattle, WA, 1999), chapter XVII.

6 Wohl, *A Passion for Wings*, p. 253. According to Le Corbusier, who witnessed the flight, Lambert's altitude was 300 metres. However, since the Eiffel Tower is 324 metres in height, Scott's perspective places Lambert about 100 metres higher.

7 Pilâtre de Rozier and his co-pilot, Pierre Romain, became the first people reported to lose their lives in a manned flight, when Rozier's balloon exploded during an attempt to cross the English Channel in 1785: *www.spartacus.schoolnet.co.uk/AVrozier.htm* [accessed 10 January 2005].

8 H. G. Wells, *The War in the Air* (London, 1907), chapter 3 [www.online-literature.com/wellshg/warinair/3/, accessed 10 January 2005].

9 Jules Verne, *Five Weeks in a Balloon* (Paris, 1865), trans. William Lackland (1869), chapter VIII [*www.intratext.com/x/ENG2061.HTM*, accessed 18 January 2005].

10 Ibid., chapter XIII.

11 Ibid., chapter XXXIX.

12 Professor Thaddeus Lowe, Official Report, Part II – Series iii – Volume iii [s# 124], *Correspondence, Orders, Reports and Returns of the Union Authorities* (1 January–31 December 1863) –# 12. *www.civilwarhome.com/loweor2.htm* [accessed 12 February 2005].

13 Santos-Dumont, quoted in Wohl, *A Passion for Wings*, p. 39.

14 Wells, *The War in the Air*, chapter 6.

15 The figure given for the number of passengers transported on DELAG airships varies between 33,000 and 40,000, depending on the source.

16 John Stroud, *Passenger Aircraft and their Interiors, 1910–2006* (Newcastle upon Tyne, 2002), pp. 5–6

17 Garros, quoted in Wohl, *A Passion for Wings*, p. 208.

18 Boelcke, quoted in Wohl, *A Passion for Wings*, p. 214.

19 Cecil Lewis, *Sagittarius Rising* (London, 1977), p. 81.

20 Wohl, *A Passion for Wings*, p. 285.

21 Antoine de Saint-Exupéry, *Flight to Arras* (London, 1942), p. 20.

22 Lieutenant J. Parker Van Zandt, 'Looking Down on Europe', *National Geographic Magazine*, XLVII/3 (March 1925), pp. 265–6.

23 Kenneth Hudson and Julian Pettifer, *Diamonds in the Sky* (London, 1979), pp. 20–21.

24 Charles Lindbergh became one of the greatest and most tireless promoters of 'air mindedness' following his celebrated transatlantic flight in 1927. He travelled the world in his single-seat plane, *The Spirit of St Louis*, demonstrating its capabilities and meeting those who could advance the development of aviation in countries around the world, but particularly in Central and South America, where he paved the way for the establishment of Pan American Airways. During his two-month tour of the Caribbean, Lindbergh met 'twelve presidents, four governors, besides many mayors, commanding generals, and other dignitaries'. This article is a diary of that journey. Col. Charles A. Lindbergh, 'To Bogotá and Back by Air', *National Geographic Magazine*, LIII/5 (May 1928), pp. 529–601.

25 Hudson and Pettifer, *Diamonds in the Sky*, p. 55.

26  Gilbert Grosvenor, 'Flying', *National Geographic Magazine*, LXIII/5 (May 1933), p. 630.
27  Lane Wallace, 'When the World Got Small', *http://www.microsoft.com/games/flight-simulator/fs2004_vega.asp* [accessed 26 September 2005].
28  Barry Schiff, 'The Flying Boat', AOPA *Pilot*, XLV/9 (September 2003), pp. 62–9.
29  'I Get a Kick Out of You', *Anything Goes*, music and lyrics by Cole Porter (USA, 1934).
30  Grosvenor, 'Flying', p. 625.
31  Hudson and Pettifer, *Diamonds in the Sky*, p. 84.
32  Versailles Treaty restrictions on German civil aviation were finally lifted in 1925.
33  Alicia Momsen Miller, *From Rio to Akron aboard the Graf Zeppelin, 1933*: *http://home.earthlink.net/~nbrass1/zepp/zepp1.htm* [accessed 25 January 2005].
34  Arthur Koestler, *Frühe Emporung* (Vienna and Zurich, 1970), vol. I, p. 2951; quoted in Claude Lichtenstein and Franz Engler, ed., *Streamlined: A Metaphor For Progress* (Baden, n. d.), pp. 119–20.
35  Donald Nijboer, *Cockpit: An Illustrated History of World War II Aircraft Interiors* (Erin, Ontario, 1998), p. 88.
36  Ibid.
37  See *http://history.acusd.edu/gen/filmnotes/memphisbelle.html* [accessed 27 January 2005]
38  Channel 4 Television, *Bomber Crew* (broadcast UK, 20 December 2004).
39  Nijboer, *Cockpit*, p. 50.
40  Channel 4, *Bomber Crew*.
41  Nijboer, *Cockpit*, p. 60.
42  Ibid. The author, Jeffrey Ethell, was killed while test flying a Lightning for *Flight* magazine in July 1997.
43  For further reading on the varied roles of the C-47, see A. Tusa and J. Tusa, *Berlin Airlift* (London, 1998), and A. Russlee Chandler III, 'Sitting Ducks over Normandy', *Aviation History* (July 2004), pp. 22–30, and in the same issue D. Smith, 'Enduring Heritage', pp. 12–18.
44  Charles Lindbergh, *Flight and Life* (New York, 1948), pp. 6–7.
45  'Flying Wing' designs were pioneered by the engineer Jack Northrop in the late 1940s. For more information, see Ted Coleman and Robert Wenkam, *Jack Northrop and the Flying Wing: The Story Behind the Stealth Bomber* (St Paul, MN, 1988).
46  Greg Goebel, 'The Northrop Grumman B-2 Spirit Stealth Bomber', v. 1.0.0, 1 December 2003: *http://www.vectorsite.net/avb2.html* [accessed 29 January 2005].
47  Hudson and Pettifer, *Diamonds in the Sky*, p. 72.
48  Giedion, *Mechanization Takes Command*, p. 466.
49  Johnson and Hibbard were members of the same design team responsible for the dramatic shape of the Lockheed P-38 Lightning interceptor.
50  Roger Bilstein, quoted in Phil Patton, 'The Connie' (2002–3): *http://www.philpatton.com/constell.html* [accessed 31 January 2005].
51  Henry Dreyfuss, 'Tailoring the Product to Fit', *Industrial Design*, VII/6 (June 1960), pp. 68–81.
52  Hughes-Stanton, *Transport Design*, p. 39.
53  Ibid., p. 41.
54  'Industrial Design at Pan Am', *Industrial Design*, VI/3 (March 1959), pp. 30–41.
55  *Airplane*, dir. Jim Abrahams, David and Jerry Zucker (USA, 1980). This was a parody of the film *Airport 1975* (dir. Jack Smight, USA, 1974), in which the stewardess lands a jumbo jet single-handedly after the flight crew are killed in a mid-air collision between the airliner and a private light plane.

56 Clive Irving, *Wide Body* (London, 1993), p. 55.
57 Ibid., p. 239.
58 Ibid., p. 241.
59 Jennifer Coutts Clay, *Jetliner Cabins* (London, 2003), p. 17.
60 *Barbarella*, dir. Roger Vadim, production design by Mario Garbuglia (France/Italy, 1968).
61 Wolfe, *The Right Stuff*, pp. 268–72.
62 Ibid.
63 Ibid.
64 Raymond Loewy, *Industrial Design* (London and Boston, MA, 1971), p. 205
65 Klaus Brauer, Boeing's interior specialist, quoted 17 November 2003: *http://www.boeing.com/news/releases/2003/q4/nr_031117g.html* [accessed 3 February 2005).
66 John Macarthy, *Excess Baggage*, BBC Radio 4, 1 April 2006.

## Conclusion

1 J. C. Crotts and W. Fred van Raaij, *Economic Psychology of Travel and Tourism* (Stroud, 1994).
2 See *http://flatrock.org.nz/topics/flying/a_two_seater.htm* [accessed 12 April 2006].
3 See *http://northumbria.ac.uk/sd/academic/scd/news/263561?view=Standard&news=archive* [accessed 12 April 2006].

# SELECT BIBLIOGRAPHY

Amory, Cleveland, *The Last Resorts* (New York, 1952)

Armi, C. Edson, *The Art of American Car Design* (University Park, PA, and London, 1988)

Baker, Carlos, *Ernest Hemingway: A Life Story* (New York, 1970)

Barker, T. C., and M. Robbins, *A History of London Transport*, vol. 1 (London, 1963)

Barthes, Roland, *Mythologies* (London, 1973)

Baudelaire, Charles, *The Painter of Modern Life and Other Essays* (New York, 1964)

Bayley, Stephen, *Sex, Drink and Fast Cars* (London and Boston, MA, 1986)

—, *Harley Earl* (London, 1990)

Baynes, Ken, and Francis Pugh, *The Art of the Engineer* (Guildford, 1981)

Bel Geddes, Norman, *Horizons* (New York, 1932)

Coleman, Ted, and Robert Wenkam, *Jack Northrop and the Flying Wing: The Story behind the Stealth Bomber* (St Paul, MN, 1988)

*Correspondence, Orders, Reports and Returns of the Union Authorities* (1 January–31 December 1863); O.R., Iii—Volume Iii [S# 124], —# 12: http://www.civilwarhome.com/loweor2.htm

Coutts Clay, Jennifer, *Jetliner Cabins* (London, 2003)

Cromley, Elizabeth Collins, 'Sleeping Around: A History of American Beds and Bedrooms', The Second Banham Memorial Lecture, *Journal of Design History*, III/1 (1990), pp. 1–17.

DAAB, *Yacht Interiors* (Cologne, 2005)

Damase, Jacques, *Carriages, Pleasures and Treasures* (London, 1968)

De Botton, Alain, *The Art of Travel* (New York, 2002)

Delius, Peter, and Jacek Slaski, *Airline Design* (Düsseldorf, 2005)

Dickens, Charles, *American Notes* (London, 1842): www.bibliomania.com/

Ehrenberg, Ilya, *The Life of the Automobile* (London, 1985)

Flink, James, *The Car Culture* (Cambridge, MA, 1975)

Forty, Adrian, *Objects of Desire* (London, 1986)

Fritscher, Jack, *Titanic* (San Francisco, 1999)

Gartman, David, *Auto Opium: A Social History of American Automobile Design* (London and New York, 1994)

Giedion, Siegfried, *Mechanization Takes Command* (Oxford, 1975)

Glancey, Jonathan, *The Train: An Illustrated History* (New York, 2005)

Green, Oliver, and Jeremy Rewse-Davies, *Designed for London: 150 Years of Transport Design* (London, 1995)

Gregory, Alexis, *The Golden Age of Travel, 1880–1939* (London, 1998)

Heskett, John, *Design in Germany, 1870–1918* (London, 1986)

Hudson, Kenneth, and Julian Pettifer, *Diamonds in the Sky* (London, 1979)

Hughes-Stanton, Corin, *Transport Design* (London, 1967)

*Illustrated Catalogue of the Great Exhibition, London 1851* (reprinted London, 1970)

Irving, Clive, *Wide Body* (London, 1993)

Kepes, Gyorgy, *Sign, Image and Symbol* (London, 1966)

Kilpatrick, Mell, ed. Jennifer Dumas, *Car Crashes and Other Sad Stories* (Cologne, n. d).

Kist, J. R., *Daumier: Eyewitness of an Epoch* (London, 1976)

Kumar, Amba, *Stately Progress: Royal Train Travel since 1840* (London, 1997)

Le Corbusier, *The Decorative Art of Today* (London, 1987)

—, *Towards a New Architecture* (London, 1982)

Lewis, Cecil, *Sagittarius Rising* (London, 1977)

Lichtenstein, Claude, and Franz Engler, eds, *Streamlined: A Metaphor for Progress: The Esthetics of Minimized Drag* (Zurich, 1990)

Lindbergh, Charles, *Flight and Life* (New York, 1948)

Loewy, Raymond, *Industrial Design* (London and Boston, MA, 1979)

Lovegrove, Keith, *Railway, Identity, Design and Culture* (London, 2004)

Lucie-Smith, Edward, *Furniture: A Concise History* (London, 1979)

Lynes, Russell, *The Tastemakers* (New York, 1954)

Marling, Karal Ann, *As Seen on TV: The Visual Culture of Everyday Life in the 1950s* (Cambridge, MA, 1994)

McCluskie, Tom, *Anatomy of the Titanic* (London, 1998)

Nijboer, Donald, *Cockpit: An Illustrated History of World War II Aircraft Interiors* (Erin, Ontario, 1998)

Pascoe, David, *Aircraft* (London, 2003)

Quartermaine, Peter, and Peter Bruce, *Cruise, Identity, Design and Culture* (London, 2006)

Saint-Exupéry, Antoine de, *Flight to Arras* (London, 1942)

Schonberger, Angela, ed., *Raymond Loewy: Pioneer of American Industrial Design* (Berlin and New York, 1990)

Silk, Gerald, *Automobile and Culture* (New York and Los Angeles, 1984)

Steinbeck, John, *Cannery Row* (London, 1958)

Stroud, John, *Passenger Aircraft and Their Interiors, 1910–2006* (Newcastle upon Tyne, 2002)

Tarkington, Booth, *The Magnificent Ambersons* (Garden City, NY, 1918)

Thorne, Martha, *Modern Trains and Splendid Stations: Architecture, Design and Rail Travel for the Twenty-first Century* (Chicago, 2001)

Tryon, W. S., *My Native Land: Life in America, 1790–1870* (Chicago, 1961)

Tumminelli, Paolo, *Boat Design: Classic and New Motor Boats* (Düsseldorf, 2005)

Verne, Jules, *Five Weeks in a Balloon* (1865), trans. William Lackland (London, 1869): www.intratext.com/x/eng2061.htm

Virilio, Paul, *The Aesthetics of Disappearance* (New York, 1991)

Vitra Design Museum, *Airworld* (Weil am Rhein, 2004)

Vivian, Charles E., *A History of Aeronautics* (Seattle, WA, 1999)

Votolato, Gregory, *American Design in the Twentieth Century* (Manchester, 1998)

Waugh, Evelyn, *Brideshead Revisited* (London, 1945)

Wells, H. G., *The War in the Air* (Chapter 6): www.online-literature.com/wellshg/warinair/6/

Wohl, Robert, *A Passion for Wings: Aviation and the Western Imagination, 1908–1918* (New Haven, CT, 1994)

Wolfe, Tom, *The Right Stuff* (London, 1981)

Wollen, Peter, and Joe Kerr, *Autopia: Cars and Culture* (London, 2002)

# ACKNOWLEDGEMENTS

I would like to thank the following for advice, information, encouragement and support during the writing of this book, help including research tips, advice on the text, imagery, technical rescue and even piloting for aerial photography: John Andrews, Dolores Barchi, Jonathan Bell, Polly Binns, Rachel Campbell, Helena Chance, Vivian Constantinopoulos, Peter Cornish, Greer Crawley, Robin Day, Benjamin De Haan, Diana Drummond, Carl Falck, Simon Forster, Paul Jarvis, Simon Lynes, Margaret Mathias, Alfonso Pons, David Rowell, Peter Slater, Anthony Slocock, Paul Springer, Clare Taylor, Damon Taylor, Arthur Votolato, Anne Weallans.

# PHOTO ACKNOWLEDGEMENTS

The author and publishers wish to express their thanks to the following sources of illustrative material and/or permission to reproduce it.

Photos courtesy of Airbus: p. 17 (top, middle); photo courtesy of Airstream Europe p. 74 (top); photos courtesy of the author: pp. 18, 22, 24, 28, 56, 64, 66, 68, 69, 72, 74 (middle), 80 (middle, foot), 87, 92, 93, 94, 100, 113, 115, 125, 201 (foot); photos The Bridgeman Art Library (© Southampton City Art Gallery): p. 35; photos British Airways Archive and Museum Collection, Heathrow Airport: pp. 8 (top), 10, 13, 170, 171, 196, 200, 209; British Motor Industry Heritage Trust: p. 8 (foot); photo courtesy of the Chrysler Corporation: p. 86; photos reproduced thanks to Citroen DS World: p. 80 (middle, foot); photo courtesy of Consumer Reports: p. 84; photo courtesy of Chris Craft: p. 143 (right); photo courtesy of Robin Day: p. 153; photo courtesy of Suzanne Fagence-Cooper: p. 148; photo Mike Fizer: p. 175; photo Lee Funell, courtesy of Ross Lovegrove: p. 216; photo courtesy General Motors Corporation: p. 96; photo indexstock.com, reproduced courtesy of Chrysler Corp. (R.P. Kingston Collection): p. 77 (foot); photo: Mell Kilpatrick (Jennifer Dumas Collection): p. 82; photo courtesy of Kyushu Railway Co.: p. 59; photos Library of Congress, Washington, DC, Prints and Photographs Division: pp. 6 (LC-USZC4-5383), 20 (Historic American Buildings Survey HABS DC, WASH, 644-1), 105 (LC-USZ62-2558); photos courtesy of NASA: pp. 17 (foot), 214; Lilian and Dieter Noack Collection: pp. 34, 160; photo Pan American Airways: p. 201 (middle); photo courtesy of Dan Patterson: p. 187 (lower right); photo courtesy of Recaro: p. 89; photo Rex Features: p. 130 (foot) (132516F); photo courtesy of Riva Yachts: p. 143 (left); photo Al Schoepp: p. 15; photo courtesy of SNCF: p. 57; photo courtesy Toyota: p. 95 (foot); US National Ocean Service (Office of Coast Survey), reproduced courtesy of the US Department of Commerce: p. 98; photos courtesy of the USAF: pp. 165, 189, 192, 193; photo J. Vack, courtesy Marc Newson: p. 95 (top); photo courtesy of Agnieszka Waleczek: p. 150; photo courtesy of Wally Yachts: p. 125; photo R.J. Welch: p. 112; illustration by Karl Zimmermann, courtesy Western Pacific Railroad: p. 52.

# INDEX